The Decomposition of Sociology

The
Decomposition
of Sociology

IRVING LOUIS HOROWITZ

New York   Oxford
OXFORD UNIVERSITY PRESS
1993

To the memory of
William "Bud" Maxwell McCord,
1930–1992

———————

Oxford University Press

Oxford   New York   Toronto
Delhi   Bombay   Calcutta   Madras   Karachi
Kuala Lumpur   Singapore   Hong Kong   Tokyo
Nairobi   Dar es Salaam   Cape Town
Melbourne   Auckland   Madrid

and associated companies in
Berlin   Ibadan

Copyright © 1993 by Irving Louis Horowitz

Published by Oxford University Press, Inc.,
200 Madison Avenue, New York, New York 10016

Oxford is a registered trademark of Oxford University Press

Library of Congress Cataloging-in-Publication Data
Horowitz, Irving Louis.
The decomposition of sociology / Irving Louis Horowitz.
p.   cm.
Includes bibliographical references and index.
ISBN 0-19-507316-9
1. Sociology—Study and teaching—United States.
2. Sociology—United States—History.
I. Title.   HM47.U6H67   1993
301'.0973—dc20   92-36400

2 4 6 8 9 7 5 3 1

Printed in the United States of America
on acid-free paper

Contents

The Decomposition of Sociology

Introduction

It is more a matter of pain than pride to have felt the need to write such a book as sits before you. The title alone, *The Decomposition of Sociology*, is a long way from the title and hopes of my work but a short quarter of a century ago. At that time, I edited a volume entitled *The New Sociology* and a few years thereafter a collection of my papers presumptively called *Professing Sociology*. The reader is entitled to ask: what happened, and to whom? Is this "sea change" a real change in the functioning of a discipline gone sour, or has the author simply lost his way, gone sour on the discipline, as it were, and in the process drowned?

Part of the answer resides in the state of the field as reported by Barbara Kantrowitz in a February 1992 *Newsweek* article on the crisis in academic sociology.[1] The sociology departments at Washington University in St. Louis and Rochester University have, as the euphemism would have it, "shut their doors." More astonishing yet, the oldest sociology department in the United States, that of Yale University, is being slashed by 40 percent through what might be called "death by attrition." And most recently, this same pattern can be observed in the decision of San Diego State University to shut down its sociology department for budgetary considerations. What makes this latest development so important is that the decomposition process of the field has now penetrated to teaching no less than research institutions. Thus, the subject of my work is hardly confined to the realm of ideological chatter.

3

The number of students earning bachelor degrees in 1991 was only 14,393, or less than half the record high of 35,996 in 1973. But this hardly begins to tell the bizarre story of departments in which professors outnumber active graduate students. Perhaps the most ironic feature of Kantrowitz's story was her selection of five titles in sociology that "offer insight into the behavior of people in groups." Of these, Emile Durkheim's *Suicide* was published in 1897 and Max Weber's *Protestant Ethic and the Rise of Capitalism* was published in 1904. She was able to muster only three more examples since that point in time; that is, for the past seventy years! While a variety of sociological newsletters and house organs reacted in outrage, little effort at factual refutation was made. Indeed, Ellen K. Coughlin, in the *Chronicle of Higher Education*, repeated and deepened this set of concerns about the status of sociology. Her series of interviews with defenders and critics of the faith alike led her to believe that sociology was in the grip of a crisis or, at the least, "a sense of vulnerability."[2] Even those who claimed only a bright future for sociology sounded as if they were delivering an autopsy.

Another general concern over the malaise in sociology was expressed by the economist Thomas Sowell, who observed that while estimates of probabilities are the shared concern of the social sciences and have an impact on every aspect of life from banking to college admissions, "let these estimates conflict with the zeitgeist, and the zeitgeist wins every time." And if this is true for politicians, judges, and the intelligentsia in general, there is scant doubt that it is doubly so for sociologists—the self-appointed keepers of the ideological flame of "partisan science."[3]

In light of what is now apparently common wisdom, I have come to the conclusion that the solution to the problem of the decomposition of the discipline does not reside with individual pique but with the unique constituencies and intellectual antecedents of the discipline as such. Further, as David Marsland in the United Kingdom has begun to show in his own brief examination of "the preconceptions and prejudices of the sociological establishment," the unraveling of a discipline is hardly confined to any single nation.[4]

What I have tried to do in this volume is provide a framework of analysis that makes sense of the apocalyptic title and gives substance to the claims registered. For I firmly believe that a great discipline has turned sour if not rancid. And it is incumbent upon those who have spent a lifetime within the mansions of sociology to offer explanations for this tragic condition. For even if one can adopt a philosophic stance with regard to the life and death of disciplines and fields of research, the occurrence of the inevitable does not make it less difficult to accept.

Let me emphasize that this is not a plea for a special kind of sociology, one that offers placebos in place of criticism; nor is this volume an assertion that the condition of a single discipline epitomizes the state of social research as such. Rather, the argument is that sociology, as a result of a special set of historical and current situations, and internal pulls no less than external pushes, has become so enmeshed in the politics of advocacy and the ideology of self-righteousness that it is simply unaware of, much less able to respond to, new conditions in the scientific as well as social environments in which it finds itself.

Indeed, such concerns are enlarged by a lingering suspicion that what has taken place was not inevitable but a function of cowardice and retreat on the part of the leaders of sociology, those people who knew better but preferred silence to resistance, collapse to struggle. The smart set was convinced that the age of ideology within sociology could be "end run" by hiding out in techniques and methodologies or, at the other end, in tepid talk about civilizational rises and falls, or gatherings celebrating past achievements, that gave offense to none but the lonely hearts.

Let me be emphatic: we are not dealing with the collapse of social research. As I state repeatedly and with absolute conviction, the condition of social science in America has never been healthier. We are witnessing a proliferation of fields and a new vitality in older disciplines that only point up with telling precision the unique condition of the decomposition of sociology. Indeed, were this part of a general malaise, then all one could really claim is that sociology was caught up in a maelstrom, in a swirl of currents that collapsed social research as such.

There have been such periods in twentieth-century history: Nazi Germany, Fascist Italy, and Communist Russia all banned sociology or severely curbed the exercise of social research as an autonomous event. In such conditions, sociology was either in the vanguard of honest, democratic effort, and its practitioners often went into exile; or it fell prey, like other disciplines, to the blandishments of totalitarian regimes—never to recover, except to receive the obloquy and condemnation such commitment to the "positive" values of such regimes deserved.[5]

In the current decomposition, sociology suffers an internal laceration. It is not the repressive machinery of government but the fanatic ideologists of the anti-statists (as least of the capitalist states) who deliver the final blow upon themselves. The core resistance of the fanatics is to the impulse to honest research, resistance to the belief that objectivity, even if never quite realized in fact, is an ideal and a value that must be held sacred at least in theory. In such a context, sociology comes to perform the politically correct roles that few other elements in society still deem worthy. Homicide is a tragedy that healthy people can recover from. Suicide is a tragedy of a much different order—it disarms the innocent and penalizes the living.

That is why the problem of the decomposition of sociology in the democratic West is a far different and more dangerous phenomenon than the destruction of sociology in earlier totalitarian regimes. As it turns out, sociology was able to survive, even thrive, in post-Hitler Germany and made a substantial comeback in post-Stalin Russia. A certain buildup of interest emerged during the repressive years. But in the democratic West, where sociology has now become identified with what Paul Hollander refers to as the anti-American spirit, no such resilience is possible.[6] Hence, prospects for intellectual recovery, at the macroscopic level, are hardly promising.

I am aware that such heavy conclusions may invite derision and disbelief. This is, after all, a work which concerns careers and jobs, no less than professional ideologies. If the bad news is the deterioration of a discipline, the good news is that—as a result of the enormous expansion of the social sciences as a whole—career and job opportunities have actually expanded. What is being described is the contraction of a field of learning, not the passing of civil society. I will be emphasizing this point in a variety of ways throughout. For my deepest concerns are not with pointing out the sad situation in one area of study but with highlighting the life-giving opportunities the decomposition of sociology holds for the professional community and the larger public.

Nonetheless, it is with the lingering hope, however dim, that decomposition can be followed by recomposition rather than death that this volume is offered to the marketplace of ideas. It will take an enormous act of courage by the leaders of the discipline and, even more, a capacity for renewal on the part of the young to reverse present trends. But if determinism is part of the baggage of outmoded and outdated sociologies, the spirit of innovation is part of the health of the scientific community. One hopes against hope that this window of opportunity will be taken advantage of. If it is, this project will have been well worth the time and effort; if not, it will simply be one more last will and testament to the death of a tradition.

Part I

THE DECOMPOSITION OF SOCIOLOGY

Science and conscience cannot in the future be permitted to develop in detachment.

John A. Ryle*

* From *Changing Disciplines: Lectures on the History, Method and Motives of Social Pathology*. London and New York: Oxford University Press, 1948.

1

The Decomposition of Sociology

The purpose of this opening chapter is to describe and explain an anomaly: in a period when the fields of social science—from the policy sciences to social planning, from public administration to demography—are expanding, one area, sociology, with a distinguished lineage and tradition, is suffering hard times. Indeed, sociology threatens to join phrenology as a pseudoscience and to share the fate of occult studies in being viewed more as a privileged language of dedicated elites than as a field of investigation broadly reflective of public needs.

Even social *ad hoc* committees of the American Sociological Association have admitted, in the words of Randall Collins, that "we have lost all coherence as a discipline; we are breaking up into a conglomerate of specialties, each going its own way and with none too high regard for each other."[1] The gaps between pure and applied research or between quantitative and qualitative theories (and many other reasons) have been adduced to explain this apparent malaise.[2] I propose a less esoteric, and admittedly more political, reading of the situation.

First, however, I want to state my belief that the social sciences as a whole are buoyant, resilient, and have a bright future. Even such traditional areas as economics and psychology have experienced great expansion, despite ideological tumult and organizational disarray. Clearly, we are witnessing no general retreat to a prescientific approach to social issues.

9

We are witnessing the decomposition of one field, sociology. Although identified, in the popular imagination at least, as largely an American phenomenon, sociology came to fruition in the relatively open political climate that existed until 1933 in Europe. At that point, Nazism prevailed in Germany and Bolshevism hardened in Russia. To the degree that sociology fought the good fight against tyrannies of the mind, left and right, it thrived as a discipline. The great names of the field—Emile Durkheim, Max Weber, Karl Mannheim, Robert Michels, Ferdinand Tönnies, Georg Simmel, and countless others—inspired both the scientific research agenda and the search for democratic options. Indeed, nearly all of the great figures in sociological history appreciated the link between free thought and free people. They may have differed on the details, but they were unswerving on the broad picture of this relationship of politics to science.

The decomposition of sociology began when this great tradition became subject to ideological thinking, and an inferior tradition surfaced in the wake of totalitarian triumphs. With Giovanni Gentile in Fascist Italy, Karl Schmitt in Nazi Germany, and Nikolay Bukharin in Communist Russia, sociology became attached to the aims and needs of the state. These sociologists saw the state as the sublime and beautiful dispenser of goods and services to the people; opposition to the state came to be viewed as antipopulist. Since sociology had always stood with the great abstraction known as the People, it now came to stand with the State. And since science had been reinterpreted as determining what people need and planning for such needs, a scientific sociology also came to be identified with the state. The state alone, it seems, could dispense the goods and services required to ensure the health and well-being of the people.

Some years later in America, the same illusion was nourished, and sociology came to the same dismal abyss into which it had fallen in war-torn and revolutionary Europe. This occurred for two reasons. First, the American experience centered on problem solving and helping people. But, in the absence of a great tradition linking science and democracy, American sociology opted for a remedial vision of social welfare—a vision that took for granted the need for social improvement while giving scant attention to the sources of social strife. Second, the inferior European tradition of sociology emerged from nineteenth-century idealist philosophy, with its goals of global domination and metaphysical certainty. And when the American academy accepted sociology as a science, it came with this metaphysical baggage. Under the influence of these two forces, American sociology could not fend off its irrational potentialities.

The idealist tradition of sociology emphasized the sources of error rather than the grounds of truth. In particular, the sociology of knowledge—which in its European form mediates the claims of the metaphysical and the empirical, the subject and the object of investigation—substituted sociological explanation or ideological critique for the rational assessment of claims to know. Thus, even before it settled its place in the scientific cosmos, sociology denied the possibility of that cosmos.

In prewar Germany, Nazism became a mode of sociological thought; whereas in the Soviet Union, Marxism was made the exclusive science of society. The Soviets affirmed sociology, but only as a partisan and class activity; the Fascists affirmed it solely as a function of race, ethnicity, and religion—a science of society, rooted in a pseudoscience of inherited biogenic characteristics.

That such basic philosophical positions were hostile to sociology as an autonomous science, and indeed denied the very notion of autonomy in the social sciences, seems to have been ignored. When not ignored, these self-destructive tendencies were incorporated into the subject by those who saw sociology merely as a tool to muckrake, a means for criticizing the status quo. Therefore, the currently fashionable attack on Western civilization and on the rationalist presumptions of Western scholarship does not encompass an external assault on sociology. Instead, such attacks have become the sociological style *par excellence.*[3]

The Soviet Union was the first nation to convert sociological theory into scientific practice. The prestige of science was harnessed to the planning machinery of the state. The price of this conversion is well known and well discussed in the literature. The machinery of repression rather than the methodology of sociology came to prevail. Even the term "sociology" was soon in disrepute. The science of society became equated with Marxism pure and simple. It did so under conditions of material backwardness, political absolutism, and intellectual atrophy of an extraordinary variety.[4]

Yet, none of this was a deterrent. Sociologists came to view the very backwardness and despotism of the Russian inheritance as a challenge. The Soviet Union was simply a screen on which sociology would project its great messages of planning for all, participation by few. Instead of providing a note of caution, backwardness and tyranny were too often viewed as special opportunities to get beyond the decadence of the bourgeois state with its deformed emphasis on opulence and selfishness.

Thus willing to delude themselves, some Western sociologists maintained the fiction of the liberating potential of Marxism–Leninism. They did so despite overwhelming evidence of human repression, including the carnage of the best and the brightest. Western sociology, far from being in the forefront of criticism, absorbed in the most tragic, mimetic fashion possible all the slogans about the partisanship of science and the class character of knowledge and its possible use in human liberation. True enough, class gave way to gender or race or ethnicity, but the myth of a single variable explaining all human behavior was a fixed idea that could not be shaken. It became embedded in the dogma of liberation sociology.[5]

As a result, during the long period of Nazi collapse and Soviet atrophy, roughly from 1945 to 1989, sociology found itself in a cul-de-sac: the flight of real talent to other forms of creative endeavor left a remnant of theorists with an ideology devoid of rational content. When the Soviet "model" utterly collapsed in the wake of the 1989 revolutions, American sociology was unable to respond. It simply fell back on domestic clichés about inherited inequities. A few grudgingly admitted that things did not work out well in Russia, but America would be different. Just give the love of sociology one more chance! Sociology was so locked into this scandalous trilogy of politicization, planning, and partisanship that it could neither back away from its dogmatic commitments nor face the rubble that was left.

Thus, sociology has largely become a repository of discontent, a gathering of individuals who have special agendas, from gay and lesbian rights to liberation theology. But this has accelerated the decomposition process. Any notion of a common democratic culture or a universal scientific base has become suspect. Ideologists masked as sociologists attack it as a dangerous form of bourgeois objectivism or, worse, as an imperialist pretension. In this climate, sociology has lost meaning apart from its ideological roots and pseudoscientific posturing. That which sociology once did best of all, support the humanistic disciplines in accurately studying conditions of the present to make the future a trifle better, is lost. Only the revolutionary past and the beatific future are seen as fit for study, now that the aim of sociology has become to retool human nature and effect a systematic overhaul of society.

A secondary meaning of "decomposition" is the separation of a substance into its elements. This has clearly accompanied the decay of sociology as a field of study. The consequence of the influx of ideologists and special interests has been the outflow of scientists, of those for whom the study of society is an empirical discipline, serving, at most, those policy planners interested in piecemeal reform. Sociology has seen the departure of urbanologists, social planners, demographers, criminologists, penologists, hospital administrators, international development specialists—in short, the entire range of scholars for whom social science is linked to public policy.

Those in this diaspora differ widely in their levels of scholarly achievement and in the social needs they address, but they share a basic shift in primary professional loyalties and identifications from a single discipline to a multitude of practical or problem-oriented specialties. Thus spawned, the new disciplines relate theory to practice in ways previously unheard of in sociology departments. Most began by emphasizing application and only later developed a theoretical perspective.[6]

The newer disciplines draw practitioners from diverse fields. For example, the majority of demographers have degrees in economics and sociology; others have backgrounds in nutrition and information systems. But, as the field of demography matures, it is developing a core of people who are being trained in and firmly identify with demography as a discipline. If we examine *Demography*, the official journal of the Population Association of America, for 1990–91, we see that slightly more than 50 percent of its authors have backgrounds in economics and sociology; others have backgrounds in nutrition, communications, and information; but most—roughly 30 percent by my admittedly informal count—are trained specifically in demography.

In a field like criminology, the situation is far more advanced. Criminology research institutes are found at most major institutions; the plethora of organizations speaks to the mushrooming of criminology in America. The *Encyclopedia of Associations, 1990,* the twenty-fourth edition of this organizational bible, shows no fewer than 225 listings under crime, criminal justice, and criminology, and this does not even include penology. Yet at these institutions, sociologists play a minor role, eclipsed by the expertise of police officers, legal and paralegal personnel, and so on. Under such circumstances, sociology is reduced to barking from the sidelines with such shrill treatises as *Against Criminology*.[7]

A similar pattern is found in urban affairs. Housing experts, architects, physical planners, environmental protection personnel, and economists drawn from business schools join with sociologists interested in the problems of housing in relation to issues of race and class. Again, we see an expansion of organizational forms, albeit not quite as wide as in criminology. The *Encyclopedia of Associations, 1990* lists approximately two hundred entries under the categories of urban development, urban policy, urban affairs, urban and regional studies, and urban studies. If we were to include relevant forms of environmental research, this number would be larger still.

In these new disciplines, we also see variations on a familiar theme. For example, many schools of public planning and housing policy are formulating their own ideologies, invariably beginning with capitalism's inability to solve the housing crisis and ending with assurances that a new social-welfare mix is the answer. That the greatest amount of social planning takes place in free societies, while housing is in utter decay in the so-called socialist world, does not seem to faze this segment of the urban-policy profession. Their radicalism is undampened by current events, and they retain barely a memory of the more traditional, and more relevant, models of social science.

It would be a serious mistake to attribute such movements and cleavages to the decay of any one discipline. Clearly, many applied fields are driven by funding sources that, in turn, are mandated by communities, states, and agencies with urgent needs for information and for practical measures to solve specific problems. The place in this process of general changes in scholarship and the philosophy of the social sciences is a separate issue, one to which I now turn.

While a return to positivism might appear the easiest way for sociology to heal itself, positivism's very lack of—sometimes denial of—larger meanings and moorings can be counterproductive. We live in a period when the false option of crude empiricism competes with varieties of abstracted grand theory for the souls of disciplines already emptied of human content. And this is a problem that all the social sciences face in common.

The politicization of the academy has also had a deleterious influence on sociology. Sociology has had a splendid opportunity to study the fundamental factors in the evolution of social systems. Indeed, between 1945 and 1965 sociologists inspired major studies of everything from the armed forces to the character of philanthropy. Sociology also found itself serving as a catalyst for changes in race relations, gender studies, and community analysis. But increasingly, sociologists thus engaged have been assaulted for findings out of step with orthodoxy and tarred by the suggestion that their work is reformist and remedial and, hence, wrong.

The revolutionary wing in postwar America, like the reactionary wing in prewar Germany, has the same hard-core, fanatical belief that remedial efforts are doomed and that only revolution—or the more frequently used word, "insurgency"—can be a proper mission. Fundamental research has been renamed "critical research." As Paul Hollander has observed, this, in turn, is simply a euphemism for anti-Americanism and anti-Westernism.[8] Sociology courses and texts have become repositories of urgings for revolution—celebrations of disparities rather than serious efforts to remove them. Every indicator of race, class, and gender differences is used to prompt the subtle or overt conclusion that American society is not only incapable of solving its own problems, but that it is the problem. As a result, sociology's impulse for remediation has dwindled. Sociology has led the charge—not only against proposed remedies to social ills but also against funding agencies and foundations dedicated to remedial goals.

A curious consequence of the radical movements of the 1960s—the displacement of reason by passion—is modeled on the Marxist demonologies that are called methodologies. Thus, the sociological has come to be defined in terms of strict political loyalties rather than scientific canons. A finely honed example of such fanaticism is the way the self-styled radical sociologist Martin Oppenheimer in the 1990s sums up the 1960s. Despite the utter collapse of the Communist empire, Mr. Oppenheimer "manages to remain cautiously optimistic about the long haul." Indeed, he says flatly that

> Marxism has become a legitimate, if not exactly respectable (fortunately) part of academia. The professoriat, the ultimate aristocracy of labor, is becoming unionized. Some of us have developed solid bridges to progressive communities outside the university, and to causes of many kinds. Our networks of comradeship are more or less intact. Although there have been a handful of renegades, and some others have retreated into privatized lives and esoteric scholarship, most of us have stayed the course.[9]

The inflation of rhetoric aside, what is most disconcerting in this passage is the definition of a presumed science in terms of comrades and renegades, progressive causes and "privatized" lives. The psychological properties of this presumed political vanguard within sociology have become the core of a discipline emptied of content. For what in Oppenheimer's remarks might seem to be his idiosyncratic fantasy or merely a hyperbolic expression has, in fact, become the common coin of the sociological realm. To see this, we need only scan the letters to the editors of *Contemporary Sociology*, the review medium of the American Sociological Association. The structure of these debates, their literary manners and psychological properties, reveal a scientific bankruptcy that makes science itself marginal to the pursuit of sociology.

What permits sociology to survive, even in its truncated form, is not its content but the fact that its practitioners remain lodged and tenured in colleges and universities. For it is here, in the last medieval retreat from the aches and pains of everyday life, where the case nexus does not prevail, that ultrarevolutionary doctrines can be foisted on the unsuspecting young with impunity. As David P. Bryden put it so simply and yet elegantly:

> In practice, courses about class and power usually attract radical teachers and repel moderates. The reading lists are predominantly if not exclusively anti-bourgeois and statist. Although in a superficial, formal sense a required course on race, class, and gender might enhance curricular diversity, in a substantive, political sense it would simply reinforce leftist points of view that are already dominant in the universities.[10]

Every disparity between ghetto and suburb is proof that capitalism is sick. Every statistic concerning increases in homicide and suicide demonstrates the decadence of America or, better, resistance to America. Every child born out of wedlock is proof that "the system" has spun out of control. Yet those who maintain this absurdity attack any effort at birth control, thus forging an unholy alliance of extremes between individuals who see increased misery as the human battering ram of tomorrow's revolutions and those who see sex education and birth control as a violation of divine law.

Sociology is now an ideology (or at least a set of ideologies) instead of what it had been in an earlier time, a *study* of ideology. Under such circumstances, the flight of serious scholarship and scientific research from sociology is inevitable. As Brigitte Berger reminds us, when sociology entered the counterculture it left us with "a matrix of consciousness in which the liberal values of rationality, hard work, efficiency, reliance upon structures typical of the modern university and the entire Western world itself, appeared as intolerable evils to be destroyed."[11] With such an approach, the fate of sociology has been placed in jeopardy: it has entered what Virgil calls "the swamp of Styx by which the gods take oath."

The swamp is not merely external. Unlike the earlier emphasis on class—which admittedly had its own problems—the newer emphasis on race, ethnicity, and gender has served to highlight the process of Balkanization within sociology. For these newer varieties of a unicausal approach demand not only greater attention but also insist upon exclusive devotion. There is a strongly implied belief that only by membership in the ascribed group can one really write about the group. Particularity becomes routine.

A sampling of new and recent journals (not restricted to sociology, it must be said) indicate the contagion effect. Thus, we now have *Feminist Issues, Hispanic Journal of Behavioral Sciences, The Black Sociologist, The Journal of Black Psychology, The Armenian Review, Contemporary Jewry,* and a myriad of others. It is not that these journals are uniformly poor, although a case could be made for far greater quality control in many of them, rather that the line between scholarship and partisanship has been virtually extinguished. To be sure, commitment is an implied precondition to writing for such journals.

Under pressure from special interest groups within academic life, issues of journals catering to them have now appeared. Thus, the entire Winter 1993 issue of *Alternatives* is devoted to "Feminists Write International Relations"—in which not the concerns of women, as such, are critical, but the centrality of what one contributor calls "Feminist Lenses." Even some highly reputable older journals, such as *Social Science Quarterly,* regularly turn over their pages to the mantra of ethnicity, race, and gender.

Sociology is thus a residue of what it once was. Its core is no longer theories of society patiently built up from empirical investigations; instead, it consists of crude caricatures of society. Sociology has become a series of demands for correct politics rather than a set of studies of social culture. Theoretical differences are evaporating as gentle intimidation displaces intellectual inquiry, and the result is an advanced form of decay disguised but not removed by the plethora of ideologists who have invaded this once omnibus "science of society."

It might well be asked: if such a dreary scenario is accurate, how can the American Sociological Association boast its distinguished recent lineage of presidents—such people as James Coleman, Stanley Lieberson, and Seymour Martin Lipset? In a curious way, however, these leaders illustrate rather than contradict my position. To begin with, they represent an earlier generation for whom a classical vision had meaning. They are also distinguished by a great distance—in practice and theory—from the organizational framework of the sociology profession today. In short, they display what *was* rather than what *is* the core of the discipline. They are the last hurrah.

There are many nuances in this decomposition. It is an uneven process that moves by fits and starts. The deterioration of sociology is not the end of social science research. It is, in fact, a signal to all practitioners of science, physical and biological as well as social. A failure to attend to first principles can result in a failure of nerve—which, in turn, means the end of science as such and the victory of ideology—a far cry from the presumed "end of ideology" promised in the heady post–World War II decades. So far as sociology is concerned, the century ends not with a bang but with a whimper.

A problem yet unmentioned is that sociology's malaise has left all the social sciences vulnerable to pure positivism—to an empiricism lacking any theoretical basis. In such a bloodless environment, the flight from sociology can inspire a flight from social science. Talented individuals who might, in an earlier time, have gone into sociology are seeking intellectual stimulation in business, law, the natural sciences, and even creative writing; this drains sociology at least of much needed potential talent.

In other words, though it has not by any means destroyed social research, the decomposition of sociology threatens to eviscerate it. This situation can be avoided only with great difficulty—by repairing sociology or appealing to the idealists and humanitarians still in sociology. This is hardly a simple assignment for those in search of a better future. It also lacks pleasant closure. But out of this intellectual rubble we must begin to face the tasks of a new millennium—I trust with more success than our predecessors enjoyed.

One example of the problem of decomposition is the breakdown of any notion of a core discipline. Increasingly, each discipline spawns subdisciplines, and these, in turn, define the total context of the scientific enterprise. Questions arise as to the degree of interchangeability of learning and teaching.[12] Nowhere is this more evident than with sociology. Indeed, a critical source of its weakness has been the degree to which the discipline has spun off areas that were formerly part of its canonical offerings.

Fields such as demography, criminology, policy research, and urban affairs (just to name a few) that were part and parcel of sociology have not only forged independent bases of operation but have also been compelled, in the process, to define themselves over against sociology. To be sure, this is not done directly. Rather, in the act of becoming "applied sciences" they have been forced to reckon with their origins.

In an area like criminology, one finds a field thriving precisely as crime concerns become greater, not lesser. The exponential growth of the field, now boasting five organizational groups at national levels and an infinite number of subfields of its own—extending from public safety to new forms of punishment—has led it to see sociology as a drag on its work. As sociology increasingly adopted a "critical" posture, it also became stridently anti-American. It became suspicious of meliorative efforts as such. This is scarcely new, since the earlier departure of areas like child welfare and social work were greeted by sociology with contemptuous delight—as if being rid of the "helping disciplines" actually strengthened the purity of sociology.

This overall attitude of contempt derived from a deep theoretical tradition that the task of sociology was to change no less than understand society. However, comprehension and change sat uneasily together upon the sociological throne. For as the century wore on, the decision increasingly was made to favor change over comprehension. There soon was no reality that theory could not overcome. Even the collapse of communism hardly caused a ripple within an American sociology convinced that some variant of social welfare or socialism was a necessary precondition for a good sociology.

There can be no denial that rates of vandalism, fire, and destruction of property are higher in some parts of a city or a nation than in others. But let a bank make a decision that the risks are too high to gamble on loans based on vulnerable property and they will be accused of redlining. Let an auto insurance or rental agency charge differential rates because of the potential of theft or damage, and the same rejection of estimates will occur. As a result, with sociology again in the lead, evidence is rejected if it contradicts the value system of the revolutionary vanguard.

Having manufactured the ideology by which decisions are made, sociologists are rarely called upon to produce evidence meeting the same standards as those demanded of advocates or defenders of the differential patterns of behavior based on actuarial charts. In this way, the sociological community is at loggerheads with the business community. And it remains so with a certain gusto that serves only to confirm traditional patterns of animus.

But as this process of theoretical triumph over empirical reality grew, so too did the departure from sociology of those who still felt that reality was worth holding on to. Indeed, the liberal tradition that nourished sociology was increasingly found in the helping disciplines, in those areas of the social sciences that sustained a belief that remedies were available for the plight of the poor and the less fortunate members of a community. The radical tradition, whether of ultra-leftist or -rightist varieties—and one could hardly tell them apart on a growing variety of issues—saw a home base in sociology.

Sociology became a home away from home for those who nourish the thought of revolution as not only an end-in-view but as an end-in-itself. It became a base for those who had no political base from whence to nourish such fond hopes of regime dismemberment. And as this process unfolded, so, too, did the disenchantment of those who entered sociology with the idea of linking criticism to construction.

The process of intellectual Balkanization soon gave way to a deeper process of antagonism: the sociologists remaining loyal to the discipline viewed application as virtual treason; while those who sought a place in the new fields spawned by sociology saw the field as useful in theoretical model construction, but not for much else. In this way, slowly, almost imperceptibly, sociology became a marginalized field, reduced to dart throwing from the sidelines of society at the end of a century that one would have thought would have seen its triumph. In field after field, warning signs of the collapse of social order were ignored. Personal reward replaced social returns as the touchstone of society.

Sociology witnessed a set of sea changes; or if one prefers, a set of begats. First, there was a demand shift: from changes in sociology to changes in society. Then there was a supply shift: the growth of new and more potent interest groups fueled sociology rather than new innovations in science and technology. This, in turn, led to a reassertion of sociology as a secular religion; this time around in the form of socialism and social-welfare models. The demand for sociology as partisanship, insurgency, and sheer property of the fanatics led to a struggle for controlling the field as a media device. What resulted was a growing disparity of proffered social-welfare solutions to a magnitude of actual social problems. Increasingly, sociology came to be out of sync with the social contexts and pressures and the tribulations confronting everything from family structure to crime networks.

The collapse of communism in Europe and Russia, coupled with continued expansion of the market societies, was the ultimate end of the revolutionary "paradigm." Not only was the prognosis profoundly wrong, but also the inability to predict with any accuracy the unfolding of the collapse of communism compromised the status of the field as a predictive instrument.[13] Again, as societies around the world rediscovered a reliance upon constitutional modes of securing change, sociology became mired in revolutionary and guerrilla modes. It hissed even louder as the world ignored its blandishments.

The shift from class to nation in actual social functioning and the rise of ethnic and racial clashes for which ideological purity simply makes no sense again caught sociology unawares. There was a subversion of sociology as analysis in favor of ideology; but worse, a subversion of sociology's historical identification with democracy in favor of a fashionable identification with destructive, irrational models of change. As such, there was a subversion of the new sociology in favor of subjectivity as the essential mode of reasoning; that is, constructing in place of examining, explaining, or predicting.

Sociology was unable to overcome the twin challenges of abstracted empiricism and grand theorizing that C. Wright Mills properly worried about. Methodological precision and moral purity became polarized expressions of the collapse of the scientific "middle." Sociology dissolved into its parts: criminology, urban studies, demography, policy analysis, social history, decision theory, and hospital and medical administration. Sociology as such was left with "pure theory": sections of itself on Marxism, feminism, and Third Worldism. It became, in short, a strident "interest group," a husk instead of a professional society.

The themes covered in the new sociology—the study of the media, administration, development, family organization, crime, and race relations—have hardly dissolved over the past thirty years. What has dissolved is the ability of sociology as such to serve as a unifying intellectual framework. In that sense the newer tendencies proved unable to transform or even galvanize the discipline of sociology; but they were able to take root in allied areas of research. What remained is the essential spirit of democracy and equity that motivated so many individuals to enter the field of sociology in the first place.

2

Disenthralling Sociology

In 1964, in part as a tribute to the late C. Wright Mills who had died two years earlier and in equal measure as an organizing device for sociologists concerned with the serious study of larger sociopolitical issues, I edited a volume entitled *The New Sociology*.[1] However favorably this compendium was reviewed, the volume had only limited immediate impact on the profession. People interested in perfect methodological exercises hardly paused to worry about, much less take seriously, the need to look at the big picture; whereas the growing legion of sociological discontents to whom the book was also aimed were taken up with movement politics of one persuasion or another and had little time for or interest in sociological work.

That event is the background for this chapter. I intend this not at all as a global reflection on social science as a whole. The worthwhile themes and purposes of the golden years of sociology have been adequately, if not fully, incorporated by the emergence of new areas of social research (for example, policy and risk analysis) and the reinvigorating of others (such as criminology, medical sociology, demography, and gerontology). Yet, it is fair to say that two decades of polarization—between unconcerned quantitativists and overly concerned qualitativists—created an intellectual environment that has provided barren soil for the survival, much less growth, of a science of sociology.[2] My purpose is to call the discipline back to its first principles by a critique of the politicization of American sociology, a process even more exaggerated and corrupt than the situation during the first two decades of the century in German university life that was described by Max Weber in his critique of Gustav Schmoller.[3]

Because it has become passé to speak ordinary truths, it is wise to recall what happens when regard for objectivity is dismissed. In Weber's words, "the task of the teacher is to serve the students with his knowledge and scientific experience and not to imprint upon them his personal political views."[4] What takes place when this task is forgotten is intellectual fanaticism, bullying of students, and subversion of learning. In such fertile soil, as the Hitlerite and Stalinist experiences demonstrate, totalitarianism thrives. Even the potential for a science of society shrivels in the face of collectivist ideologies, while the individual and the community lose substantive meaning. The strong go elsewhere, the weak acquiesce, and the field of sociology dissolves into positivist fragments servicing the state.

One of the great myths in contemporary inner critiques of sociology is the presumption that its practitioners share an unswerving allegiance to mainline methodological premises.[5] This myth argues that sociology still strives to be value-free and that its adherents are exclusively guided by a search for truth. This has been repeated often and by elite figures in the disciplines constituting the sociological core; consequently, actual superstructural conditions of social research in the mid-1980s have been obscured, even dangerously neglected.

I wish to correct these misimpressions and describe the current worldview of sociological "theory" in present-day America. The picture revealed is not pleasant; it is often painted in ideological black and white, vaguely responded to by nervously unconcerned greyish figures, without much space for the shadings and nuance that characterize the actual human condition—not to mention the social condition.

Sociology has been profoundly and differentially politicized by its practitioners. Such politicization stems from an ostensibly leftward drift at the same time that American society as a whole has rediscovered such rightward verities as moral absolutes, universal standards, and inherited values. This countercyclical drift accentuates the widespread alienation of sociology as a discipline from social life and its replacement by undistilled ideology or, at times, quite distilled theology. The hectoring, badgering tone of an Old Right has now found a home in the New Left.

At the moment, partisanship loses force in all areas of scientific research; it insinuates itself as a paradigm in the battle of sociology against impure bourgeois thought. In consequence, the ameliorative, policy role performed by sociology throughout its twentieth-century history is now weakened, perhaps beyond repair, from within.[6] What has occurred in the inner history of sociology is a split between methodologists and ideologists—between those with a neo-Kantian belief in the purity of the research process apart from policy considerations and those with a neo-Marxian belief in the purity of revolutionary goals apart from empirical considerations. The interplay of such inner strains and external demands forms the leitmotif of the politicization of sociology in America. In this formalist battle, the corrective potential of sociology is thwarted; its actual existence as a viable, liberal discipline is curtailed.

The illustrative vignettes that follow repeat themselves in one sociological pronouncement after another. The collapse of the functionalist–positivist paradigm in the late 1950s left a gaping intellectual void filled by assorted claimants to a nonexistent sociological throne. The soft pragmatism of earlier American social theory (that is, the Parsonian school) quickly fragmented. This gave way to a hard empiricist reaction to abstraction on one side and a manifest assertion of the supremacy of ideology over reality on the other. Old intellectual loyalties and even older institutional networks, such as those at Harvard, Chicago, and Columbia, were seriously ruptured.

My position directly confronts these dual-track false alternatives in sociology—what Mills long ago identified as abstracted empiricism on one side and grand theorizing on the other. Both are now sprinkled with ideological frosting. My text stays close to the professional literature, and I avoid dealing with institutional shifts and organizational commitments. These latter areas have been widely covered and hyperbolically reported in the literature on the profession. Further, my remarks are aimed less at odious constructions than at intellectual reconstructions. This is an attempt to analyze, not just describe, the outcome of the twentieth-century crisis in Western sociological theory.

The literature upon which I draw, even if modest in intellectual significance, is nonetheless substantial in raw volume. What this literature lacks in sophistication it makes up for in bully-pulpit tactics. Even to state mild opposition to this new extremism takes considerable courage. The areas to which I draw attention, while by no means all inclusive, are not entirely random. I use materials that have recently appeared in refereed journals of sociological scholarship. They typify a pattern in which extremism extends its grip on the culture of sociology at that point in time when totalitarian modes of rule are crumbling worldwide.

The pattern of manifest politicization within sociology moves incxorably along a line of reasoning that begins with a denial of the normative features of society in general and concludes with a categorical attack on the foundations of American democratic society in particular. Curiously, the utopianism of future societies is considered admissible grounds for judgment, whereas the norms of present societies are *a priori* ruled inadmissible. I offer more here in the nature of agenda setting than a complete description of the sociological situation.

The key objective of ideological extremism from the outset has been the total repudiation of the normative character of the social system. This is done by subjectivizing behavior and making structure nothing more than perceived idiosyncratic actions. The new moral relativists have informed their readers that in the wonderful world without norms there can be no deviance—only alternative life-styles, contextually situated.[7] Thus, it becomes easier to declare norms null and void than to address the issues of crime and its prevention. All behavior can be interpreted in morally neutral terms, without consideration of legal or moral institutions of marriage, family, or community. If such institutions are dealt with, it is often in terms of their structural shakiness or intellectual hypocrisy. Social solidarity is old-fashioned, whereas individual swinging, whatever its larger costs, is new-fashioned. The message is clear: life has little merit or virtue apart from action.[8] The act requires no justification beyond itself.

The one area of social science in which absolute adhesion to canons of ethical neutrality is insisted upon is the study of social deviance. The ideas of heterosexual relationships as normative, or marriage as a sacred or bonded state, are viewed as intolerable partisanship, lacking an evidentiary base. A transvaluation takes place: those formerly in the closet come out with a vengeance, and the heterosexuals come to be considered not just antiquarian but also lacking in innovative spirit or pure adventure. The revolt against cultural tradition becomes a demand for revolutionary behavior—and, not unexpectedly, a new absolutism.[9] Acquiescence to such behavior is often insisted upon at the personal level as proof of a break with the hoary past.

By mystifying the relationships between those who commit crimes (violate norms and laws) and those who are victimized by criminals, crime is liquidated as readily as deviance. Crime simply becomes an interaction ritual, a network of evil associations. Even such a new criminology does not go far enough to suit some sociological ideologues. So, they have designed a new approach which abolishes crime by fiat.[10] Crimes are disaggregated: those with victims (murder and assault upon the person) are distinguished from those with presumably no human victims (prostitution, drugs, gambling). In this manner, crime, if not abolished, is at least greatly reduced.

Homicide alone remains as a category worthy of the label "crime." Even here we have those whose celebration of the political criminal, the intelligent criminal, comes close to excusing homicidal acts. In a world without good or evil, but only crime and punishment, the ultimate evildoer, the central criminal, is the state itself. The more democratic it is, the more criminal, since hypocrisy heads the list of crimes. This is a central theme of many talented figures of the new criminology. In their intoxication with social change, the merits of social order are simply denied, at least as long as such an order is linked to capitalism.[11] The debates among the new criminologists revolve around whether crime itself is a meaningful category, rather than the collapse of normative structure in advanced societies.

Donald Black, one of the most avant-garde thinkers, has now come to consider the conduct of crime not as "being an intentional violation of a prohibition" but as the opposite, "the pursuit of justice . . . a mode of conflict management." Indeed, "viewed in relation to law, it [crime] is self-help." This is not parody but a conviction that "crime often expresses a grievance" and is thus of a piece with other forms of behavior. No effort is made to test the worthiness of the grievance. Since "the criminality of crime is defined by law" everything is up for grabs in this sociological version of Mailerism.[12] Professor Black at least had the kindness to admit "that crime and its seriousness may be explained with a theory of law as a departure from common sense."[13] Writers of sociological fiction do not bother with such demurrers.

In its politicized version, the original purpose of American society is characterized as an attempt to create a collective, cooperative community. But the fall from grace engineered by the bourgeoisie substituted the base notion of personal aggrandizement, or entrepreneurialism, for the utopianism, the place of grace.[14] Thus, the collapse of community led to the rise of repugnant individualism.[15] This, in turn, led America to the cult of hero worship, primitivism, narcissistic pleasure, and private happiness at the expense of the public good. In this example of the politicization of social research, the task of social science is to restore a sense of community through collectivization and to implement planned change to achieve the good (that is, socialist) society. The failure to achieve community is seen as subversion by a new order—one far more complete than the one replaced.[16]

Traditional liberal values are stood on their collective heads, as it turns out that every reform measure in the United States since the New Deal is doing little more than staving off the inevitable collapse of capitalism. In its purified form, welfare operates as a clever form of social control. In the period from 1960 to 1980, it was an integral part of governmental efforts to recommit a rebellious poor population to the existing social order. Welfare performs a social control function for the capitalist state. In this universe of tautological discourse, where the welfare system has succeeded it has done so by false consciousness, by deceiving people. Where it fails, the will of the people triumphs over capitalist antics.

The ability of American society to prosper in, much less survive, this welfare environment, and to be able to do so as a result of rightist rather than leftist pressure was neither anticipated nor even considered a distinct possibility. Thus, the ideologists are reduced to continuously predicting the final end of capitalism.[17] Since not even the new ideologists can overtly support terrorist attacks on the social fabric, a group of revolutionary scholars have devised methods to cleanse terror of its repugnant practices by linking it with freedom struggles or by making opponents of such linkages the creatures of state repression or social segregation. Analyses of "powerlessness" translate into effective modes of using "insurgent power," including the "tactic" of utilizing riots and terror to create a process of "tactical interaction." Under the veil of Aesopian language, the threat of gangland ("foco") styles of revolutionary behavior is approved..

In a critique of the notion of organizing the poor, some sociologists argue against helping the poor to reap the benefits of a welfare system. Instead, in manifest Leninist rhetoric, they ask and then answer: "What, then, is to be done? Our position, which is implied in our criticism of separatist organizing, is that the basic task of activists who are concerned about poverty is the promotion of socialist consciousness among the rank and file in the trade unions."[18] The response of welfare liberals to antiwelfare radicals is not reassuring: suggesting only differences in means to reach a social nirvana in which all people end up in the same place.[19] There is a taken-for-granted use of sociological periodicals to vent socialist and even communist polemics and ambitions. The assumption of the legitimacy of violence in the name of sociology is a hallmark of the new ideologues.

The announcement that elites control, manipulate, and dominate corporate life in America is most fashionably argued in terms of interlocking directorates; that is, the degree to which key individuals dominate the board rooms of every industry from boots to books and liquor to lumber. The hidden predicate is that such interlockers mastermind everything from foreign policy to fiscal policy. At times, the conspiracy is even given a special geographical hideout.[20] A belief in the existence of a unified industrial network is extended to the conclusion that there is a single unified high command in American society as a whole.

We are informed, despite evidence of a decline in centralization of business authority, that "a single, unified network" of interlocking directorates exists, with "a high degree of centrality." All contrary information notwithstanding (emergence of middle-sized businesses or diversification of the labor market), the existence of conspiratorial elites remains a fixed point of faith that "power and centralization are fused in the corporate economy."[21] How such relationships cross industrial lines, whether or not directorates actually share a common ideology, and the distinction between newer and older industries in differentiating policies all vanish before the onslaught of the interlocking directorates—an inch away from the dictatorship of the bourgeoisie (whose mythic characteristic serves to justify the real characteristic of the dictatorship of the party apparatus in the Soviet Union).

Sociological or psychological evidence to the contrary, all industries not fitting the elitist-conspiracy model, all areas not participating in the directorates, are simply dismissed. For some, American society is a cleverly rendered political conspiracy to maintain bourgeois power—now known as corporate liberalism—intact, even if and when that power is federalized.[22] Newer developments moving to decentralize and debureaucratize control are either dismissed as impossible without a deep restructuring of society or met with glum silence, as if every reform is a painful interlude needlessly postponing the joyful storm of industrial protest and class revolution.[23]

Only within pockets of sociology do the dogmas of pure Marxism burn deeply. In a tormented article, "Capitalist Resistance to the Organization of Labor Before the New Deal," we are told that "a steady structural tension, however latent it may be at any point in time, necessarily continues to exist between capital and labor because in our view, it is rooted in the very fabric of capitalism." Whatever is actually observed in labor–capital relationships, the mysterious workings of class struggles must go forth in the bowels of America. The piece concludes, predictably enough, with a statement of the New Deal "legacy" to America: "capital's disorganization of American workers as a class in the first three decades of the twentieth century."[24] Aside from the error of historical fact—the New Deal commenced only in the fourth decade of this century—is the metaphysical conceit that any effort to mitigate or meliorate the circumstances and life-styles of workers is not only economically suspect but a clever political conspiracy perpetrated by President Roosevelt and his New Deal. The corruption of this form of stratification analysis had reached a point where any evolution in Western economic systems can only be described in harshly negative and disparaging terms.

Unadulterated ideology often appears in the guise of adulterated scholarship, as in the article entitled "Eminent Domain: Land-Use Planning and the Powerless" by Alvin Mushkatel and Khalil Nakhleh. The purpose of this article (one of many of this genre) is to show that the policies of the capitalist United States and Zionist Israel are similar: to expropriate from the poor their lands and homes in an arbitrary and capricious fashion. Inner-city rebuilding in ghettos is compared to West Bank resettlement. We are assured that "both nations were able to carry out widespread programs removing large numbers of their people from their land and their homes." Within the Israeli context, this is part of "the larger Zionist goal, maintenance of a Jewish-Zionist state"; within the United States the "liquidation of Indians was a precursor to hegemonic domination of the poor."[25] That such partisanship could appear in a refereed, scholarly journal dedicated to the study of social problems is indicative of the sorry state of social analysis in America. It is also a consequence of the legitimacy endowed by the phrase "Zionism is racism" as a result of U.N. resolutions. Social scientists as partisans feel no constraints on their practical missions given such implicit sponsorship and global legitimation.

Given the nature of Israel as a "social problem," it is little wonder that political scientists, rather than sociologists, have been in charge of this area. The prototype is Ali A. Mazrui, whose two articles in *Alternatives: A Journal of World Policy* set the agenda for linking Zionism, Nazism, and apartheid. We are told that on such diverse issues as racial exclusivity and national origin, "Zionism has even more in common with pan-Germanism than does Afrikaaner nationalism." We are informed that the Jews who "ran away" from Nazism went to Palestine "to establish for themselves a 'pure Jewish State.'" The conclusions reached become axiomatic:

> By the 1980s the logic of Zionism had come full circle. The successors of Dr. Malan in South Africa, once fervent in supporting Hitler's anti-Semitism, are now in alliance with the successors of Chaim Weizmann and Ben Gurion. The strategy of displacing Palestinians has found a point in common with the strategy of stripping Xhosa of their South African citizenship. The anti-pluralist element in Zionism is now aligned with the anti-pluralist element in apartheid. Somehow the ghosts of Auschwitz in Germany, of Deir Yassin in Palestine, and of Sharpeville and Soweto in South Africa, are suddenly in bewildered communion with each other.[26]

In this entire rhetorical outburst of analogical reasoning, the historic claims of Jews to the region are never acknowledged; the condition of the Arab population in Israel as the most advanced and economically viable in the entire Arab world is never mentioned; and the Holocaust and its incomparable loss of lives under Nazism are not examined. The unique support of the Jewish communities and Israel for racial equality and political democracy in South Africa is also never mentioned. Since writing the articles in *Alternatives*, Mazrui has become more rather than less strident in his political utterances.

At this point, he considers Israel "as the most arrogant sovereign state in the world scene since Nazism." He slightly mitigates this canard by noting that Nazi Germany was "both absolutist and militaristic; but the state of Israel is only militaristic." The overall impact, the impression left by the Mazrui approach, has become a linkage between Nazism and Zionism, a middling of the democratic characteristics of Israel, and, at the same time, a muting of the totalitarian characteristics of Hitler's Germany.[27] All is bathed in a social-movement rhetoric of "God, gold and gender." Far from washing away the sins of Jews, it only serves as a forceful reminder of the ideological uses of social science.

Ideological extremism has a way of demanding holistic and totalistic solutions. World-systems theory has no difficulty sliding into an overt anti-Israeli position. One such formulation, made by Anouar Abdel-Malek, argues that "the roots of violence, the roots of global war, the road toward armament, lie in the historical structuration of the international order, *i.e.*, in the historical surplus value, beginning in the fifteenth century." Underneath such grand theorizing is a concern not for history but for "a new civilization project . . . a human and progressive *realpolitik.*" Wherein are the challenges to such a humanistic *realpolitik?* Perhaps in the rise to power of the epigones of the Trilateral Commission of the Cold War and the Zionist apparatuses at the center of the leading imperial power of our times—one that is facing the rising tide of national liberation movements and social revolutions now unfolding in the world.[28] This is the political undertow, the substructure of "a perspective which recognizes the primacy of analysis of economies over long historical time and large space, the holism of the socio-historical process, and the transitory (heuristic) nature of theories." Thus spake Immanuel Wallerstein. The steady drumbeat of the linkage of world-systems theory and anti-Semitism and anti-Zionism is made in the name not of socialism but of a partisan politics and historiography.

Having dismissed the American dream, or at least converted the dream into a nightmare, several world-systems advocates turn their minds to greater things than Zionism. They argue that the evolution of modern capitalism, of which Zionism is a pinprick, is a unified global affair, with an imperial core, a transitional semiperiphery, and a downtrodden periphery. This emerges from the assumption that there is a unified world order dominated by an imperial core—now located in Washington. In previous centuries, the core was characteristic of an epoch.[29] The hidden predicate of such world-systems theorizing is that to transform the social order, to give it its freedom, one must smash the imperialist core presently in the United States and permit the democratic socialist periphery to emerge. For the new ideologues, the destruction of the United States in its present form is the key to the liberation of the Third World.

In this scenario, the task of social science is to aid the liberation process by exposing the process of imperial domination. The only thorny problem for this group of political sociologists is what to do with Mother Russia. This dilemma is neatly resolved by making Russia part of the world capitalist system.[30] Whatever constrictions exist are a function of the way in which Russia was integrated into the world capitalist economy. Socialism as an ideal can never be sullied. The magic of words always comes to the rescue when there is a paucity of facts; or when the facts reveal a Gulag. This is less disturbing to the new ideologues than the wonderful organicism from which one can strike a blow at the imperial center and watch the world as it is transformed into something unnamed but doubtless superior to American society. Intellectual myopia is not unconnected to partisanship in sociology.

Increasingly, the form of scholarship, its paraphernalia, is retained, while the conclusions have a decreasing relationship to scientific methods. Conclusions that are uniformly and at times shamelessly ideological are proffered as social science. Michael Burawoy, writing a lead article in the *American Sociological Review*, claims that "irrespective of state interventions, there are signs that in all advanced capitalist societies hegemonic regimes are developing a despotic face. . . . In this period one can anticipate the working classes beginning to feel their collective impotence and the irreconcilability of their interests with the development of capitalism." And back to the *Communist Manifesto* for a little dash of inspiration: "The material interests of the working classes can be vouchsafed only beyond capitalism, beyond the anarchy of the market and beyond despotism in production."[31] Nary a word in this article described real despotism; nor are we informed of just what great utopia is awaiting us beyond capitalism. It takes little imagination to see what this author has in store for his society.

With the notable exception of those social scientists who possess a working familiarity with such European figures as Antonio Gramsci and Leon Trotsky (Nicos Poulantzas and Ernest Mandel come most readily to mind),[32] we have to search far and wide for any serious estimation of Soviet communism by the new ideologists.[33] The utter bankruptcy of Russia as a model for Western democracies has led to the displacement of utopia by myopia. The rationale for the denial of Russian reality is varied, but almost invariably it comes down to the claim that those who live in the United States are centrally responsible to critique their own domestic scenario. Whatever goes on, or went on, in Russia, is presumably to be taken care of by the Russians. Another formulation is that it is difficult to know what is taking place in Russia, so why burden critique of America with fatuous comparisons to a foreign body. The same inhibitions do not keep such analysts from doing violence to empirical conditions or from commenting on the rest of the world.

What prevails are lengthy discourses on Marxism, socialism, radicalism; any abstraction ending in "ism" that is removed from the realities of Russian power. In this manner the heady wine of theory is restored to its pristine primacy. In eastern Europe discussions of Marxism and socialism are relegated to the world of black humor; but in the United States, with notable exceptions, the sacred texts of yesteryear are pored over with a sobriety and affectation that could well inspire religious zealots. The movement from a new sociology to a new ideology is, in effect, a shift from an evenhanded critical analysis of power, authority, oppression, and exploitation—wherever it manifests itself and under whatever auspices—to a thinly disguised anti-Americanism that uses the rhetoric of social science to express animosities that would otherwise be quickly challenged or readily repudiated.[34] This sad outcome of the critical tradition of power analysis must be stated frankly as unacceptable if any hope is left for a restoration of scientific principles in the present sociological environment.

The difficulty with this nearly total silence on Russia and, at times, continuing rationalization of its past totalitarian system as a function of the dominant world capitalist market is that such a species of social science is placed at odds with the common wisdom.[35] An overall negativism comes into force that has neither a mass base of support nor a critical elite upon which to draw. "Theory" becomes a refuge from reality and hence emptied of the critical contents which adherents seek to begin with. The inability or unwillingness to examine political and social systems concretely and comparatively is the Achilles' heel of the new sociological partisans.

The new politicization of sociology is an attempt to ideologically redefine the tasks as well as the operations of research. Not only the general theory but also the specific practice of sociology is at stake. What starts as pluralistic definitions of sociology ends as monolithic politics. Jean-Bernard-Léon Foucault showed us how historically the definition of madness was used to broaden the range of what is considered sanity or normality. Those who fail standardized skills or intelligence tests can seek to improve their skills and pass the tests; or, as is increasingly the case, they can seek to change the standards by seeking waivers or changes in the tests. Homosexual behavior in the early part of this century was viewed as a basic illness, a failure to make the transition from childhood to adulthood. Now it is increasingly considered an alternative life-style, having nothing whatsoever to do with infantile regression or childhood fantasies. We can multiply examples of the use of differences in definitions to reach policy alternatives.

At one level the issue of politicization in sociology deals not with social science as such but with cultural metaphor. The problem becomes how to translate such cultural metaphors into political practices. Seen in this light, the rhetorical transformation of sociology into a constant barrage of ideological beliefs is part of an ongoing struggle for the minds of the masses. For the devotees of partisanship as truth, sociology has simply become a turf, a terrain, on which such ideological struggles are to be carried out.

The legacy of Communist-bloc nations is that, in the past, they routinely entered international social-science activities with a fierce dedication to ideological persuasions. They justified attending global gatherings by mapping strategies, disbursing literature, defining the enemy, and holding internal caucuses on the premises of sociological gatherings. Ultimately, they captured executive councils that control social-science associations and in this way came to define sociological agendas.[36] These tasks have now been absorbed by the "vanguard" thinkers within the advanced nations. This is but a singular illustration of how social science emerges as metaphorical struggle which, in turn, has an impact upon the nature of political alignments. If this seems to be a harsh mapping of ideological life on the darker side of social science, it is only because confrontation with such an inhospitable reality invites some form of response. At the root of the response is escapism through empiricism—that is, the assertion that micro-study is uniquely rewarding and, hence, uniquely beyond the ridicule of colleagues.

In parody of intellectual pluralism, the field of sociology boasts (in addition to the now mandatory section on women, the black caucus, and a cluster of other minorities) a variety of organizational shapes and forms, some directly sponsored and others benignly tolerated. These range from associations for the sociological study of Jewry and the Society for Christian Sociology at one pole to gay and lesbian sociological caucuses at the other. In between are such specialized concerns as political sociology, Marxist sociology, and medical sociology that are converted readily enough from intellectual pursuits into interest groups—each with its own institutional rules, regulations, leaders, and members. In such an organizational climate even the potential for intellectual renewal or synthesis becomes unlikely. The call for the unity of science, even of one science, is viewed as a means to subvert the newfound organizational autonomy of these specialized areas. That real science is mocked by such bureaucratic procedures dare not be an issue raised, or else these "specialities" will take their few hundred members out of the parent organization, which may already be reeling under the shock of membership reduction. Thus, what began as an intellectual parody ends as a professional charade, with the science of sociology the main loser.

A typical example of mixing and thoroughly confusing science and ideology is the treatment of women. For a segment of organizational life, women have now become an interest group, a caucus within the body of social science. In an entirely typical program, the Southwestern Social Science Association lists as fields of study economics, geography, history, political science, social work and sociology, and the "Caucus for Women." Quite apart from the extrapolation of the sex/gender variable in this entirely aprioristic and illicit way, such isolation, such ghettoization, serves to remove the study of sex and gender from the everyday operations and purview of the major disciplines. In this organizational caricature, scholars come to fear talking about, much less removing, such assaults on science. The empirical essence of looking and seeing which variables explain what processes and structures is sacrificed in favor of special privileges and prerequisites to whatever core group claims yelling rights. What began as a righteous rebellion against benign neglect has now become an absolute ban against the spirit of inquiry as open-ended, color-blind, or gender-neutral.

In the attempt to promote new forms of the democratic, to deny no one a fair hearing, sociology has become a simulated replica of real-world confusions, rather than an analytical tool for explaining or predicting social behavior. *Opera seria* turns into *opera buffa*. The dissolution of a scientific core moves lockstep with the dismemberment of the organizational loyalty *per se*. The social-science organizations are afraid, timid, or simply unable to say no to any group that can claim five or more members. In their attempts to forestall dwindling attendance, they create the reverse effect: a breakdown of organizational loyalties accompanies the dissolution of intellectual coherence. The multiplication of this phenomenon in every major branch of sociology has the global consequence of casting doubt not only on one field of study but also, sadly and mistakenly, on social science as a whole.

The metaphysical presuppositions of the politicization of sociology represents a major conversion of an "is" into an "ought." Social research is historically laden with biases, prejudices, and wrongheadedness—often caused by the elevation of a single variable into an apriority. In the past it was race, class, even nationality and geography. In our times it is more likely to be sex, gender, race (this time with reverse moral intent), or age. But the goal of social research, which is to understand and transmit the facts of social life truthfully, has been transformed into the eternality of ideology, the inevitable sociology of knowledge distortions which result from the biases of the investigator.

In part this is a consequence of a social-science legacy in which theory emerged out of philosophical tradition rather than empirical observation. As a result, the social and the scientific have had a one-hundred-year history of uneasy alliance. This dualism has served to keep alive the ideological embers of nineteenth-century doctrines long after they have run out of intellectual steam and long after they have had much relevance to actual events in Europe, much less in America. In the absence of normative convictions, historical antecedents come to fill this gaping epistemic void; but such antecedents come with the baggage of every European dogma from Marx to Mosca. What was at one point brilliant observation performed with limited tools of research is turned on itself; it is transformed into derivative theory aimed at frustrating the use of scientific method or, when that is not feasible, at least limiting such usage.

Whether by design or accident, the end product is nothing other than the subversion of sociology as both method and theory. Ritual incantation displaces experience and experiment alike. Typical is a recent piece entitled "It's Good Enough for Science. But Is It Good Enough for Social Action?" in which the author, Ditta Bartels, concludes by saying, "It must be recognized generally that scientific knowledge is not objective, and hence cannot serve as the undisputed rock-hard base upon which potentially discriminatory political decisions may be erected." While we are never told on what, then, to base our judgments or recommendations, "analysts in the history and sociology of science" who have a unique divination into the "social nature of science" are instructed to "share this understanding with scientists, with politicians, with unionists."[37] The strange thought that conflicts among scientists, politicians, and unionists may be what sociologists should be studying never enters the thought processes of sociologists like Ditta Bartels. What we receive instead is a none-too-subtle resurfacing of the totalitarian argument concerning the "partisan" character of science, the son of the earlier "class" or "race" character of science. Ideology, then, moves sociology from studying the social sources of distortions to celebrating bias in the self-declared political vanguard.

Instead of reacting to this sort of insurgent approach which curiously reflects the despairing conservative vision that "social science itself is the very disease it proposes to cure,"[38] it is more fruitful to argue that the impact of social science on the moral thinking of America is highly varied and forward looking. The level to which ideology has come to permeate and penetrate a given discipline is the measure of disease, undermining a sense of real progress. By such a criterion, sociology is a uniquely troubled discipline precisely because it has subverted, more than any other social-science discipline, the methods of science and the process of inquiry in favor of an aprioristic intellectual table thumping. The call to disenthrall sociology is but a first step to recall the discipline to its great and good first principles.

My emphasis on the ultra-leftist effort to co-opt sociology should not be interpreted as an endorsement for an ultra-conservative vision. There is no such option in social research. The rage for order and the faith in tradition—hallmarks of conservative ideology—remain at loggerheads with the practice of social analysis, with a sense of the present determining research agendas. Looking backwards no more than longing for utopias is what sociology examines, not what it apes. The implicit reductionism of the New Left is scarcely improved upon by the explicit moralism of the New Right. Looking for new gods to replace failed gods is neither strictly nor loosely speaking a function of social research. Sociology can survive only to the extent that it retains its links with the great traditions of the social science as a whole. When sociology breaks ranks with social science, it also breaks faith with the empirical constraints nourishing this discipline. While a full-fledged analysis of sociology and the conservative tradition is beyond the scope of this chapter, we need to recognize that the future of sociology, as its past, is bound up with the scientific tradition and not ideological travesty. However difficult and tortuous this path may be, it is the only one that preserves a discipline first from distortion and finally from decimation.

What has been lost in this transmigration of facts into values is the essential aim of social research: to overcome and move beyond partisanship in the social and human realms of interaction. It might well be that the goal of a value-free social science is unattainable. Along with others, I have consistently argued against a strict functional-structural explanation that would naively deny bias in everything from sampling to topic selection. Having said this, having raised an awareness of the special forms of bias lurking in the world of social life, we are still compelled to aim for the same canons of objectivity in the social sciences as are presumed to be present in the natural sciences. The politicization of social research has led to the celebration of inherited weaknesses at the expense of research strengths in sociology. Therein lies the ideological rationalization for the current uses of sociology as a vanguard of revolutionary politics. Therein lies the pending demise of sociology as a vanguard discipline in the social sciences.

What is so fascinating about the ultra-left position is its assault upon liberal values as a set of operational guidelines for sociological work—and that includes an appreciation for the varieties of possible explanations. Roach and Roach speak of a "retreat from class" as zealots speak of a fall from grace. Whom do the new ideologues attack? Why, the new sociologists—that is, "the mélange of Kap-statists, one or another variant of structuralists, critical theorists, and spongy eclectics."[39] There is a lesson in this for those who think for a moment that they will be protected from assault by the thin veneer of radical chic. The hard core is not easily deceived on such grave matters as ideological perfection.

We live in unusual sociological times. Many seem to be put upon, claiming to be in a sorry or beleaguered state. The kernel of truth beneath these self-serving claims of being in a minority condition is that the paradigms that governed the field in the past have been sundered with a thumping finality. All concerns in "Disenthralling Sociology" are less statistical issues as to who is in the mainstream or in the margins and more what makes the extreme politicization of a social science to the health and survival of minorities and majorities alike. Whether one is a social-welfare advocate or a free-marketeer is not central to my argument. What interests me is the defining role of extremist ideologies in the distortion of sociological work.

This chapter addressed certain tendencies in sociology and was not an effort to provide a capsule review of the Bolshevik Revolution or of its painfully modest gains. This is a complex subject in its own right. But what is the legacy of Marxism–Leninism in Russia that was, which is being worn as a badge of honor by radical segments within Western sociology? Rather than burden such people with my personal views, I cite from a highly reliable assessment of Russian social science leveled by sociologist Tatyana I. Zaslavskaya. Her hands-on assessment, made in 1987, before the collapse of the Soviet empire, says all that needs to be adduced in support of my fear that the politicization of sociology leads to devastating outcomes for the life of the field. The fall of Soviet communism only confirms Zaslavskaya's line of analysis.

> Let's say it right out: social studies in this country have hardly been in the vanguard of society. They have rather been bringing up the rear. . . . We hold one of the last places among the developed countries in the level of social statistics. I do not mean the raw data being kept in the closed files of the Government's Central Statistical Administration, but the results of data processing that are then openly published and

become accessible to the public at large. There has been a real downturn in this respect since the second half of the 1970s. The publication of census data has become skimpier since the 1959 count. More and more categories of social information are being kept from social scientists. Among data that are not being published in the Soviet Union are statistics on the distribution of crime, on the frequency of suicide, the level of consumption of alcohol and drugs, the condition of the environment in various cities and areas. . . . As a result sociology in our country is at a much lower level than, say, in Poland or in Hungary, not to speak of the developed capitalist countries."[40]

My view is not an attempt to turn the clock back, to retreat into the void of sociological positivism or criminological primitivism. Neither is it to seek solutions in obscurantist rhetoric and authoritarian doctrines which promulgate everything and explain little. It is to assert the supremacy of sociological explanation over all forms of ideological explanation. If class explains less and less over time, then its use as a central variable will decline—demands for ideological purification and revolutionary allegiance notwithstanding. This is no less true for other organizing variables: race, gender, age, occupation, ethnicity, religion, or whatever. The source of sociological strength is to use whatever elements in the social text that are available to critical challenge and public scrutiny. In this sense, the fate of sociology is inextricably linked to that of democratic societies. On that bedrock does my case for the disenthrallment of sociology come to rest.

3

Sociology and Subjectivity

The common denominator of the social sciences (evident in sociology in particular) is subjectivism, the tendency to deny that human behavior is normally driven by a reasoned response to a knowable reality. I do not dispute the role of subjective factors in decision making, or even the place of human will in directing the course of events; I deny that the subjective element obliterates all objective elements. To say that the world has no rules in the name of ideology is no better than to offer a diet of strict determinism.

Many would agree that criminals "work the system," but that does not mean that the criminal-justice system is a figment of the criminal's imagination. Nor would anyone seriously reject the notion that people construct, invent, manage, manipulate, and market their individual worlds. But it is a big jump from recognition of the subjective—or as it is commonly, although not always accurately, known, the symbolic—to denial of an external world of commonly shared experience.

We might have hoped that at this point in the philosophy of science, such an obvious restatement of epistemology would be superfluous; but this is not so. Indeed, those who in the past were most vociferous in asserting the paramount claims of the "real world" are now found touting the claims of the subjective—so much for the hoary idea that science and naturalism are unique preserves of "progressive humanity."

In the previous chapter, I sought to explore the emergence of a contemporary sociology as pure critique of present society. In a variety of settings—criminology, deviance, stratification, among others—I showed how social research had become hostage to ideology. Characterization had become reduced to caricature. High crime rates were seen only as an expression of capitalist disintegration, and criminal behavior became a covert expression of revolutionary action. Deviant behavior was simply a term of moral opprobrium; all social norms were really bourgeois norms; and opposition to such norms represented alternative life-styles at the least and revolutionary consciousness at the most.

In one area after another, this same formulaic thinking became the rage: urban studies was the study of how capitalist society makes decent planning impossible; medical sociology was the study of how greed pervades the behavior of professionals and how the system prevents full and adequate coverage of medical needs of the poor and the elderly; and international development was dependency theory, a thinly or thickly veiled resurrection of the Leninist doctrine that all forms of backwardness are a direct consequence of the colonial and imperial experience. All questions had one answer: the evil of capitalism. All problems had one solution: the good of socialism.

Social theory as crude political reductionism had its moment in the sun, but that moment is passing. Amid a sense of embarrassment that utopian appeals do not convince the "broad masses" at which they are aimed, a rebellion in the ranks has emerged because such approaches are intellectually unproductive. They are thoughtless and aprioristic and, hence, feeble forms of analysis and poor tools for prediction. Widespread recognition has taken place that such forms of impassioned apriority are unlikely to attract serious, much less the best, new students, and they indeed threaten the existence of social science as an autonomous set of disciplines. Such belated awareness of academic vulgarity has taken on groundswell proportions and represents a positive response to the enthrallment of sociology as a revolutionary ideology. But more sophisticated varieties of this fanaticism have developed in an effort to preserve ideological thinking.

For shorthand purposes, I will call this new form of enthrallment "sociological subjectivism," although it is by no means a disease from which political scientists or social historians have escaped. This radical subjectivity has its roots in the brilliant awareness that the process of social interaction is also an expression of the uneven distribution of wealth, power, and status. In terms Erving Goffman might use, it came to be understood that mental patients "work" the psychiatrists in order to gain release from institutions; criminals "cool out" the "marks" in order to plea bargain for shorter sentences by the courts; women employ their cunning and charms to gain ascendancy in a male-dominated work environment; and poor nations offer themselves as proxies to wealthy nations in exchange for support and favors.

An entire movement that came under the rubric of symbolic interaction tapped a rich vein of qualitative research. From George Herbert Mead through Herbert Blumer, it cut through the sentimentality of conventional varieties of radical and reactionary ideologies to show not just the dull face of oppression but the many faces of collective response to oppression. In so doing, symbolic interaction opened up powerful research possibilities of a flintier sort than anything possible within the straitjacket of conventional reductionist social research.

In its most sophisticated form, the interactional perspective made devastatingly clear that all human beings come to the table with a mixture of motives and a set of assets no less than liabilities. Welfare mothers are not powerless in confronting the welfare establishment. The homeless are not without resources in gaining public support, even from the wealthy. Language minorities know how to make majorities uncomfortable within specific enclaves and contexts. In short, this perspective seemed to provide a way out of the political morass of over-simplified analysis, without at the same time surrendering a critical or democratic cutting edge.

Curiously, symbolic interaction did not have the cathartic impact its founders imagined or its followers hoped to secure. For as it became clear that under the scalpel of the masters of theory, interaction cut both ways— top down and bottom up—the doctrine of symbolic interaction was left open to the charge of relativism. What the more radical subjectivists now aim at is absolutism—at least a sense that history is inexorable, irreversible, and will simultaneously stamp out capitalism and replace that world system with socialism. As a result, and over time, the interaction was replaced by the action that is the force of the clever wit or the brutal will. The dialogue was replaced by the hustle. With this sleight of hand, the democratic impulses of symbolic interaction were undermined and sabotaged, to be replaced by an activist, often incoherent, doctrine of change with strong authoritarian claims on its advocates.

The power of sociological subjectivism requires a deeper look at the enthrallment process in social science. What we have is a far more powerful intellectual tool, yet one that has taken social science down the same road of irrationality and normlessness as did older forms of ideological reasoning within the social-research process. I would even argue that sociological subjectivism is but the latest version of a century-old revolt against reason—one that in the wake of failure of radical movements to take root, much less take hold, in America and Europe becomes more strident and more insistent in its claims. It does so precisely at a period of time when some semblance of a return to reason is a feasible development in social science.

It is best to start with actual history and conclude with pure ideology, on the proper Marxian supposition that what starts as tragedy ends as comedy, in scholarship no less than in dynasties. The rhetoric is slightly different for each discipline; but the uses of social research to explain all forms of unease and discontent as a consequence of the larger social system will be evident.

In a quite decent book, *The Invention of George Washington*, Paul K. Longmere argues for a demythologized version of our first president.[1] George Washington emerges here, quite properly, as someone with "prodigious energies in a continuous quest for honor, for validation of his society, person, and public character." The author concludes by noting that at his core, Washington was a man with "a desire for distinction, a yearning for public esteem that ultimately became a quest for historical immortality." The problem with this demythologized view of Washington, "as an astounding performance prodded by a mixture of egotism and patriotism, selfishness and public mindedness" that is a "spur of fame," is that we end up with little new knowledge or information on Washington and much demythification of the first American president based on psychological and personality traits that are common to all who lead. The goal of a founding father devoid of heroic properties is realized; but at a cost of a larger context in which heroism was embedded in the quest for a first new nation, and not the lionization of an individual.

The notion of "invention" employed as a mode to demythologize is well within the historical and analytical tradition. Less satisfactory, as we move from history to political science, is the notion of politics as itself illusory or symbolic. A good illustration of this is the recent effort by Stanley Karnow, *In Our Image*, which develops the thesis that the background to the Vietnam war is the efforts of the great empire builders of the past.[2] In particular, rule of the Philippines made the United States a great Pacific Basin power.

As a result, this country was plunged into World War II. The Japanese military historians of the 1930s could not have formulated the history of the Pacific Basin any better. Preparing the area for American conquest, rather than Japanese militarism or adventurism, becomes the center of Karnow's attention. The arrogance of power was considered underwritten by the imaging of a powerful United States upon the powerless Philippines. Thus, as we move from George Washington to Theodore Roosevelt, we move from invention to image, but the myth-making foundations of American power are held to be central to the troubles of the world.

The political-science literature has taken subjectivism to the next higher level. If historians display ambivalence between reality and image, the subjectivist political scientists simply see the political process as a set of constructions. No one better argues this premise than Murray J. Edelman. In his recent work, *Constructing the Political Spectacle*, Edelman sees utilitarianism as the great myth itself: "For most of the human race political history has been a record of the triumph of mystification over strategies to maximize well-being. Indeed, attempts to search out a solid base of interest theory is usually so ineffective that it even buttresses hegemony by helping political elites in their construction of a stable enemy that justifies political repression."[3] For Edelman, not only is the political process an artifact, a construction, but also attempts at transcendence only solidify such constructs and place them at the disposal of unsavory elites. In this way, any semblance of mass democracy or broad political participation is seen as either chimerical or delusional in character. What starts as a critique of antidemocratic tendencies in the political process ends with a denial that even a modest democracy of self-interest can be established. In the more ambitious version of this scenario, mass culture is viewed as a creation, a construct, of mass media in which all of our values and beliefs are managed from above. Such is the message, for example, of *Creating America*, by Jan Cohn;[4] it is ostensibly a study of George Horace Lorimer and the *Saturday Evening Post*, but with far more ambitious intellectual fish to fry: the American ethos.

The notion of reality as a construction is not limited to the political arena. The new radical subjectivism has developed theories of sexuality along the same or similar lines. For example, in the excellent study *The Construction of Homosexuality*, David F. Greenberg argues with great force that assumptions of the etiological or biological basis of homosexuality simply reflect a "bias used to discredit historical figures by suggesting without sufficient independent grounds that their behavior is irrational or neurotic."[5] Greenberg's approach, while claiming to deny all bias, shows a strong bias against moral claims or societal norms. He proudly concludes that he "made no assumption that society has moral boundaries that the prohibition of homosexual activity maintains, that the society as a whole gains by the prohibition, or that the perceptions and responses we described are explained by their consequences." These are called simply "gratuitous and frequently misleading assumptions." The same courtesy is not extended to those who might claim either that sexual preferences are more than constructions or that biological or psychiatric claims have any basis in fact. Thus, without asserting a preference for homosexuality, any sociological or psychoanalytical mooring against such preferences is knocked down with this subjectivist conception of the relativity of moral judgment and empirical standards. Homosexuality as an alternate life-style becomes the only standpoint permissible.

The most recent spin on this theme is the assault on inherited moral notions. Thus, texts such as Cynthia Eagle Russett's *Sexual Science: The Victorian Construction of Womanhood* see such virtues as devotion to family, civility, and respect for law more observed in the breach than in practice in nineteenth-century England.[6] Hence, the ethos of the time was essentially a construct, not a reality. There is always a gap between observed behavior and proposed norms. So that this work, like so many others in this genre, aims to knock the props out from under any moral construct, certainly from under one viewed as less than edifying to the new woman.

In some measure, the rise of media as central to political processes has accelerated if not fostered these subjectivist claims in the political process. The spate of such articles as "Television and the Construction of Reality" is typical.[7] Ira Glasser, the executive director of the American Civil Liberties Union, contends that television is homogenizing "mainstream culture by bringing it into the American living room." Glasser goes on to say that "Any kid can turn on television and suddenly get all the sexual innuendo, all the steaminess, the political views that they don't like, the relationships that they don't like, and the values that they don't like."

But who are "they"? And, what is to be done? For the dilemmas posed by such a view of constructing, and not just mediating, reality fly in the face of conventional Marxism with its presumption of market and mass manipulation. Here, Glasser accepts manipulation as a built-in good. The magic qualities of television are seen to "create social reality." Its very seaminess, its very construction of a seamier-than-life world, becomes a good, a rallying point for the powerless and downtrodden. Thus, the conventional liberal argument against television *laissez-faire* is turned on its head, with the construction of evil as portrayed by television itself a positive force in the enlightenment of the masses. That such a view is simply a cynical view of free speech as a means to promote disaffiliation—not to describe reality—escapes Glasser's attention.

The sociologists, among those who partake of this genre, have moved from invention to construction to downright manufacturing a reality. Again, it is the media that remain or emerge as central in this new school of subjectivism. If Glasser sees the process of manufacture as revelatory, others see the same process as obfuscatory. Thus, Mark Fishman, in a widely regarded book, *Manufacturing the News*, sees the bureaucracies as prepackaging news—in press releases, policy reports, city council minutes—before disseminating it to the press.[8] By mutual consent (and here Fishman comes close to advocacy of the conspiracy theory of mass communication), reporters reduce complex happenings to simple cases that represent bureaucratic realities rather than the original, often confusing facts. Fishman claims that if the news were gathered in different ways, a different reality would emerge, one that might challenge the legitimacy of prevailing political structures. Underlying this is the idea that the United States, far from having a free press, is a place where news is manipulated by cooperating elites to the degradation of the masses. In all of this, there is not a single word on the contrast between free presses and controlled presses (such as those that exist in totalitarian and authoritarian nations throughout the world). The notion of manufacturing news or managing reality comes to stand for a denial that the United States is a place with a free press, an entity worth defending instead of destroying.

By the time we reach the ideological never-never land of Edward S. Herman and Noam Chomsky, the notion of manufacturing the news becomes one part of *Manufacturing Consent*, the title of their new book, subtitled "The Political Economy of the Mass Media."[9] Aside from the shameless effort to disguise his own past defense of the Pol Pot regime (which destroyed millions of people) by equating this with "genocide in Cambodia under a pro-American government," Chomsky, with histrionics, claims that an underlying elite consensus structures all facets of the news, and the marketplace of publishing significantly shapes the news. Everything is manufactured: the presentation of electoral processes in friendly and enemy governments, a lack of investigative zeal in Watergate, the Iran-contra hearings for failing to indict the executive branch of government, and the notion of mass media as sheer propagandist tools of the wealthy and the powerful. The theme of manufacturing as a delegitimation of the American system is again central. The notion that mass media bias may actually be in favor of liberal or radical goals is simply dismissed, and subjectivism becomes linked to a thoughtless radicalism which asserts that "the mass media of the United States are effective and powerful ideological institutions that carry out a system supportive propaganda function by reliance on market forces, internalized assumptions, and self-censorship, and without significant overt coercion." Far from a notion of mass media as responsive to mass wishes, or even part of a mindless sort of business routine, we have a notion of media control and domination that lacks even elementary credibility.

A clear expression of this sort of analysis based on subjectivist premises, and its potential for political destruction, comes from a 1984 article in the *Mid-American Review of Sociology* by Michael R. Hill, who concludes by bemoaning the absence of works on ideology—not to expose shabby thinking as a road to ruin, but rather the absolute reverse: the need for ideology as "legitimate sociological thought."[10] He concludes by saying, "It is time to turn the tide: ideology first, axiology second, epistemology third." It is a sad truth that the tide has indeed been turned, with outbursts of ideological thinking and subjectivist orientations bordering on solipsism being viewed as nothing less than the touchstones of an authentic social science.

The undiluted utopia that a narrow group of visionaries wants becomes the basis for radical critique of the existing society. This long march from historical analysis to ideological celebration of the irrational has come to displace conventional forms of Marxism that have either moved into "mainstream" analysis or been perceived as insufficiently responsible to the demand for widespread chaos and change for its own sake. An extreme example of this higher subjectivism is the work of Styephen Pfohl and a "school" of sociological thought in which not only is the line between the personal and the social obliterated, but also the line between the sociological and the fictional. It is a world in which the distinction between the fictional fantasy and the the polemical exposition is done away with. Any notion of evidence is replaced by sentiment. The struggle against heterosexism, racism, imperialism, genderism, and just about any other "ism" ever invented is wedded to such deconstructionist concepts as "the double or nothing" and "the eye/'I.'" The play of language, the madcap use of lower and upper cases, and the violation of syntactical rules replaces actual events or recent history. Radicalism is saved by appeals not to determinism or historicism but to such personal qualities as exoticism and healthy sexuality, whatever that may turn out to be. The bohemian ideology takes on the sharp edge of totalitarian adventures. And while this may impinge little on the lives of ordinary people, it impinges heavily on the current state of affairs in sociology.[11]

What starts as a perfectly reasonable expression of social-science interest in taking into account subjective elements of behavior and converting them into objective realities, ends with a denial that reality is anything other than subjective. As the century wears on, it also wears thin for those who predicted the rapid demise of industrial capitalism, the end of bourgeois democracy, and the rapid emergence of a world revolutionary movement in which only varieties of socialist or communist power would or could be debated among reasonable people.

The series of revelations in the post–World War II period— starting with the horrors of Stalinist rule and ending with the collapse of belief in the twin shibboleths of Soviet political vanguardism and economic infallibility—also played a part in this artificial fusion of the subjective and the radical. The old Marxists could say with pride that subjectivism was the last refuge of bourgeois philosophers who could not face the truth of the dying of the capitalist light; but such intellectual hubris was denied to radical sociologists who could not face the facts of a communist blight.

A Draconian bargain was struck; a temporary alliance between the subjectivists who denied the validity of social structure and the revolutionists still locked into categories of struggle and conflict against the existing social order. With a collapse of the "objective" models provided by the Soviet Union, China, eastern Europe, and Cuba, subjectivists and Marxists were linked in an uncomfortable alliance on editorial boards of journals, in interest-group caucuses, and in shared cynicism about a present world they detested and a future world turned sour.

As the cause of genuine reform no less than nationalism deepens within the former republics of the U.S.S.R., these unhappily alienated elements have made a perfunctory break with Marxism– Leninism, in the name not of probity or sobriety but of carrying out a revolution that Gorbachev in the U.S.S.R. and Deng in the People's

Republic of China presumably have betrayed. One might call this the Vietnamization of the social sciences. The problem is strategic: isolated within the United States, such a public break, while spiritually pleasing to those demanding purity of theory, has had the effect of further isolating the new subjectivists from any political support base and increasing their marginality.

This new subjective radicalism, based on the revolt against reason, is not unlike what took place in the *fin de siècle* at the close of the nineteenth century, a mystification of culture in the name of immediate gratification. Its advocates are more enraged than engaged. The *fin de millénaire*, like the *fin de siècle* one hundred years earlier, reflects the failure of the socialist dream to materialize. Indeed, the failure of socialism to emerge even as part of the political dialogue may stimulate in a precious few even more intense criticisms of society. As intellectual life begins to separate itself from social life, the universities see themselves in opposition to all other institutions and set in motion a place, like the *Bourse du Travail* (labor exchange) in the *fin de siècle*, where good fellowship, socialism, and upper-class bohemian values can flourish unimpeded and uninterrupted. In this way, secular cultism, in its isolation from the mainstream, sees its version of the social world confirmed.

At present, subjective radicalism is not a university-wide phenomenon. Such applied fields of study as business, engineering, medicine, and law are seen as agencies of the larger society on campus and, hence, defined as philistine in character. The embattled liberal arts, the social sciences, and literature, become the last repositories, if not the sole depository, for such extremist positions. Every worldly defeat, every movement in a contrary direction in the larger world, only serves to confirm the image of these new subjectivists. While the world of ordinary people invents, constructs, negotiates, and manufactures reality, our new subjectivists are unique in feeling the pulse of the people (this elitist nonsense alone is retained from the Leninist vanguard) live the good life and assert fit and proper ideological premises.

Apart for the epistemological problems involved in basing a social science upon an obliteration of any real-world structures, the Achilles' heel of this view is that the cynical disregard of popular wills and political decisions leaves untouched the broad masses for whom these elites profess such love. It is a view that underscores a sense of hatred born of frustration that popular demands are made for law and order, family and kinship, free-market economies rather than controlled-market systems—in short, appeals to some sort of objective structures that the subjectivists scornfully disregard or dismiss.

The last great wave of radical subjectivism broke up in Europe over the similar failure of intellectual elites to convince broad masses that real differences in values and structures did not exist between France, England, Austria, Russia, and Germany, between democracy and monarchy. The breakup of the present mood of irrationalism takes different forms: essentially, the abandonment by present leaders of Russia and eastern Europe of earlier forms of authoritarian ideologies. Stalinist tendencies to convert partisan claims into historical realities are no longer feasible. The willingness of these nations' current leaders to abandon ideologies in favor of pragmatic economics and ethnic politics has thrown the ideological vanguards of post-Marxian subjective radicalism into a quandary.

The essential question becomes: will this breakup of subjective and irrational tendencies take place before or after the demise of the older social sciences? This is no small matter for their practitioners. As this subjective radicalism moves into older areas, more serious social researchers are abandoning older fields and setting up new areas of work in which ideological criteria do not prevail and scientific norms are permitted and even encouraged. The invasion of obscurantism, dogmatism, and partisanship, and even their celebration, does not dampen the thirst for solid social science; but such thirst is slaked by moving special research into greener pastures. Policy analysis, public administration, criminology, demography, urban planning, communications studies, and a host of other new fields emerge as primary sources of personal identification and public performance. This leaves the new irrationalists to twist slowly in the wind. Unfortunately, as they turn, so, too, are the husks of the older disciplines desiccated.

We tend to see social-science categories and disciplines as relatively stable, if not fixed. The new wave of politicization in the academy proves that against irrational claims, no science is immune and no social scientist is safe. The struggle remains the message—in social science no less than in society at large. However, that struggle is now internal to the social-science community and no longer simply between advocates of social research and partisans of political ideology. In reviewing how well we have responded to the "manifest decline of the American civic order," Daniel Patrick Moynihan has wisely noted that the ultimate answer to subjective doctrines and partisan passions of all sorts is the simple realization that there is "a dynamic process which adjusts upwards and *downwards*. Liberals have traditionally been alert for upward redefining that does injustice to individuals. Conservatives have been correspondingly sensitive to downward redefining that weakens societal standards."[12] In this broad recognition of the principle of indeterminacy in social science, we can preserve the sense of science and, no less, the sensibility of democracy.

4

Fascism, Communism, and Social Theory

The first three chapters emphasized the inner history and traditions of sociology—intellectual moorings that strongly determined its current malaise. But sociology also operates within larger intellectual contexts derived largely from an American experience that is selectively derived and defined. In Europe, the wartime conflict between Nazi Germany and Communist Russia polarized the two varieties of totalitarianism. The polarity also made possible a choice between bad (Nazi) and good (Communist) forms of totalitarianism. This spilled over into the postwar era and, indeed, made disbelievers in the Cold War out of a substantial portion of American academics—in particular, the sociologists. With the exception of a small handful of courageous figures, sociologists managed to go through the entire cycle of 1945–91 without producing a body of comparative analysis of democratic and totalitarian social structures. That silence of the sociological lambs itself serves to define the strange impact of the larger culture.

However, despite early categorical rejection of fascist doctrines, especially in their European forms, the American social-science community was alarmingly receptive to varieties of marginal social doctrines that made their way into the confines of the social structures. A particular strain in this virus is what I have chosen to call left-wing fascism. This is not meant as a clever play on words. I am attempting to show that the sources of much sociological anti-

Americanism, or, at the least, the denial of the legitimacy of the American political system, derived from this special variety of the totalitarian temptation.

The work of Lewis Feuer provides a kaleidoscopic view of many tendencies and trends within contemporary social science. Above all, he illustrates the fission–fusion trend in intellectual currents as exemplified in the work of Spinoza, Marx, Freud, and Einstein. Not only does he write brilliantly on these figures now appended with "isms," but his work is permeated by socialist, physicalist, and psychoanalytical explanations of political events.[1] Hence, Feuer is an old-fashioned thinker in the best sense, someone who combines in his person radical persuasions with democratic practice in a variety of fields. His work is not so much anticommunist as it is antitotalitarian. I dare say, he has been singled out as a Soviet *bête noir* precisely because he has so artfully disentangled radical rhetoric from totalitarian reality.[2]

I am struck by the obviousness of this in reviewing his superb work on Marx and Engels on politics and philosophy. What emerges is an individual who believes fully and firmly that the Marxian kernel can be rescued from its Stalinist and ex-Soviet protuberance. He suggests therein that "American development is out of phase with the rest of the world. America, disenchanted with its own Marxist venture of the thirties, is learning the language of conservatism, and is finding itself ever more removed from the Asian and European worlds." I do not know if Feuer would still adhere to this severe judgment, but I am certain that he would continue to implore us "to re-learn the meaning of Marxism." Feuer well appreciated that as "freedom is reborn in Eastern Europe and Asia, it will speak in the Marxist idiom and try to disenthrall the universal humanist bearing of Marx's ideas from their Stalinist perversion."[3]

What I would like to do in this chapter is show how this process of perversion and enthrallment across generations has worked its way out as a special variety of the American social idiom. Specifically, I want to show how the phenomenon of left-wing fascism has become one way of accomplishing what Feuer terms a process of acknowledgment of that tremendous segment of reality which the Marxist philosophy has come closest to grasping.[4] To study the totalitarian phenomenon seriously means to get beyond dogmatism into actual ideological permutations and combinations which still retain a lively sense of politics.

We must first examine the social foundations of totalitarian doctrines in American life and the connection (or lack thereof) with their European counterparts. Essentially, my claim is that there is a powerful crossover between left and right, between communism and fascism, on American shores. This crossover is blurred by the exaggerated claims to uniqueness by both sides and often obscured by the different demographic backgrounds of those adhering to left and right forms of totalitarianism. In order provide some basis for this serious claim, I must examine recent American social history; in particular, those key historical figures who represented left–right linkages—sometimes knowingly, at other times inadvertently—as a function of the political context of American life at its peripheries. In so doing, I am seeking to explore the roots of political marginality in American political culture. For it is precisely the totalitarian characteristics of both extreme right and extreme left that have not only irrevocably fused them but have doomed them with respect to any conquest of political power. I hasten to introduce a caveat: this is an exploratory statement. Much more work will need to be done to prove conclusively this point of view. But it at least holds open the prospects for resolving long-standing debates not only about why America has had no "socialism" but also, no less important, about why it has resisted "fascism" at the same time.

The ideological roots of conservatism are partially located in its need to reconcile two conventionally opposed attitudes toward life: the heroic and the moral. The former is most often associated with self-expression, the latter with self-denial. The renunciation of joy in favor of destiny takes on a peculiarly Christian cast in twentieth-century conservatism. Writing of Oswald Mosley, his brilliant biographer, Robert Skidelsky, points out that "Fascism's fault, and by implication his own, as he came to see it, was too much paganism and not enough Christianity." But the seeming inability to fuse the higher men with political purpose meant that "Christian love could become the method only when the higher men are not underneath but on top. Till then it would be necessary to say with *Zarathustra* that 'love hath an end.'"[5] Thus spake Sir Oswald. We are not dealing with either a foolish position or an inconsequential individual. Even if not at the level of Friedrich Nietzsche, Mosley in his journeys through socialism, laborism, fascism, coupled with a Europe purged of American capitalism and Russian communism alike, well articulates a resurrected classical conservatism not far removed from its current expressions on both sides of the Atlantic. The question may be asked: Where do those of the Jewish persuasion fit into such a model ideology? The answer is that, clearly, they do not. And to believe that some sort of commitment to a naked anticommunism could possibly paper over a sometimes manifest, often latent, anti-Semitism was at the core of Mosley's naiveté. It distinguished a right and left in an age when the latter still gave lip service to religious and ethnic freedom. The same remains the case a half century later.

A frequently struck theme of conservatism has been the defense of Western values and Western civilization as a whole. A marvelous and wholly prototypical example is Charles A. Lindbergh, whose extraordinary technical exploits throughout his life were performed in the name of a rather ordinary ideological context in which "power to be ultimately successful, must be backed by morality; just as morality must be backed by power." As his biographer Kenneth S. Davis, succinctly notes, "the morality he referred to was synonymous with Christian ideals."[6] The advocacy of a world organization guided by Christian ethical principles was a far cry from isolationism of a secular socialism. It was no less removed from the Jewish persuasion as well.

The statements made by Lindbergh expressing the horrors of a "ruthless dictatorial system" developed in Soviet Russia, in which millions were "denied justice" and "unknown numbers" labored as slaves and by which "a record of bloodshed and oppression never equaled has been made," anticipate precisely the position outlined by Aleksandr Solzhenitsyn in *The Gulag Archipelago*—and, moreover, echo the solution he advocated. We cannot ignore the dangerous imbalance of so many conservatives in their rather tepid response to Nazism and the Holocaust in contrast to their impassioned response to communism and Babi Yar. This very absence of "even-handedness" is what makes an alliance between conservative ideology and Jewish theology a curious and temporary coalition at best, a misanthropic association at worst.[7]

An elusive form of conservatism is that tradition linking its philosophical tenets of absolute morality and political probity with a critique of capitalism. As Sheldon Marcus, the biographer of Charles Edward Coughlin, noted: "Regardless of prior ideological commitment on the left or on the right, they were attracted by Coughlin's slashing attacks against the concentration of power in the hands of big business and the federal government's passivity in the face of crisis."[8] The prevalence of historical short-sightedness being what it is, we tend to view left-wing fascism as a recent phenomenon. But the roots are deep in American populism.

Coughlin regarded Marxism as the "parent of both Nazism and communism," left and right versions of the "same bird of prey." He was an early advocate of political evenhandedness: "We regard every organization against war and fascism as a menace until the officials of such organizations incorporate the war against communism in their program." His anti-Semitism, while late in developing, was a cornerstone of his publication *Social Justice* and probably the center-piece of the message to his followers. Nor should we forget his antiwar stance, his belief in an America "self-sustained and self-contained." Not unlike the present libertarian advocates of isola-tionism, he urged others to "be ready for the defense of our country and homes from alien aggressions, but never again be the agressor."

Historians of the American Right have increasingly become aware that their formerly held neat distinctions between left and right extremist politics were convenient but largely chimerical. In point of fact, fascist doctrines in America were often tinged by popu-list critiques of big business and oligarchical concentration, while communist doctrines were just as often permeated by elitist lead-ership principles and contempt for broad sectors of working people who failed to transcend "trade union consciousness" and move into "vanguard positions." Nor were these differences exclusively be-tween left and right; they were also evident in European fascism's idea of unity versus national purpose. State control of the economy was as great (if not greater) in Hitler's Germany as in Stalin's Russia.

Emphasizing the work of Huey Long and Charles Coughlin, Alan Brinkley points out that they resonated to a different set of concerns from their presumed European mentors.

> At the center of their message was a commitment to a major shift in the locus of economic power in America, not to the state, but to small community institutions and to individual citizens. Long and Coughlin contemplated a far more fundamental assault upon the "plutocrats" and "financial despots" than the European fascists ever attempted. . . . Long and Coughlin envisioned a far different and more limited role for the government than the fascists proposed. Its purpose would not be, as in Germany and Italy, to subordinate individual economic interests to the central goals of the nation. It would rather, liberate individuals from the tyranny of the plutocrats, restore a small-scale, decentralized cap-italism that would increase, not restrict, economic independence.[9]

What we have, then, is a picture of the American Right more in tune with what is now referred to as the "new conservatism" than identifiable with the main currents of classical European fascism or communism.

In his work on the Protestant far right in America from the depression to the Cold War, Leo P. Ribuffo makes essentially the same point, underscoring the concerns of far right and far left in operational terms. He points out that "far right activists and their opponents were influenced by the fear of deception that characterized the whole inter-war period." "When the United States gets fascism," Huey Long was supposed to have said, "it will call it anti-fascism." The remark was apocryphal. The underlying premise that fascism in disguise would triumph was shared by socialists and centrist liberals, Communist theoreticians, and the editors of *Life*. Ribuffo makes the further point that "fear of the far right undermined the left's commitment to civil liberties."[10] Indeed, this is the essential thesis of his book. What I find most interesting is this new awareness that extremism, totalitarianism as such, is a key aspect of American social history. The forms of such extremism may at times be right-wing, at other times left-wing, but the core message seems unchanging; only the sources of positive utopias (that is, Germany or Russia) might be quite different. We are thus in a position to take this story up to the present. The period from 1933 to 1953, from the emergence of Hitler to the death of Stalin (if not their "isms"), is well represented in the literature. The more challenging task is to examine the decades since—both as a legacy of this ambiguous totalitarianism and in the context of current social and political events.

In taking such figures as Mosley, Lindbergh, and Coughlin as exemplars of conservatism, I certainly do not mean to imply, even remotely, that a tradition as rich, deep, and original as conservatism can be reduced to such quasi-political types. It certainly is not to identify Edmund Burke or Alexis de Tocqueville with twentieth-century demogogues. But these marginal figures of our time typify a certain configuration of ideological thought that sought to transcend major struggles of the century: between communism and capitalism, totalitarianism and egalitarianism, intervention in and isolation from international conflicts. But the lynchpin in these efforts at synthesis was appeals to absolute, dogmatic value systems that, whatever their labels—"Americanism" or "nativism" being the clearest illustration—appealed to the organizing power of Christian doctrine. Again, while it would represent little more than calumny to identify an authentic Christian mission with such rabid political theologies, the appeal of Christianity was a serviceable tool in the exclusion of Jews and other nonbelievers or infidels.

This core aspect of traditional conservative doctrine, or "left-wing fascism," is something the neo-conservatists Norman Podhoretz and Irving Kristol, among others, have had a difficult time explaining to their liberal constituencies. But even if they do not buy or sell the bundle of goods offered by the old conservatives, even if they expressly and emphatically repudiate its anti-Semitic core, they are left with parts of an intellectual machine that form a leitmotif with the older conservatism—a defense of Western civilization and a perception of that civilization as being directly and primarily threatened by world communism. The issues thus quickly move to the empirical; that is, can such an anticommunism sum up the troubles and tribulations of Western democratic societies?

Fascism has recently been called "the system of ruins."[11] In its "pure" Hitlerian form, this was certainly the case, but in its "neo" form, in its linkage of bringing an end to the social division of labor and the manifest rebellion against industrial-commercial society, fascism with its left posture, with its humanistic twinge, has taken on renewed meaning. No writer of our time has expressed these new linkages more ably than Ernst Nolte:

> Fascism can be directly compared with Marxism of the Soviet nature only in its radical form, in respect of its inner solidarity and its appeal to comrades of like mind in all countries: Italian fascism, in its phase as a development dictatorship, and more than ever the Croation Ustase and the Rumanian Iron Guard were in fact, on the contrary, more like many of today's "national liberation movements" than like late National Socialism. . . . There is nothing more grotesque than a "theory of Fascism" which denounces capitalism with much sincere indignation as the root of Fascism, at the same time overlooking that the theory identifies itself with conditions which show all the formal characteristics of Fascism. It is not astonishing that the liberal capitalist system produces Fascism under certain circumstances, but it is astonishing that in the great majority of cases Fascism has not succeeded in gaining power in spite of certain circumstances. The explanation can only lie in the fact that this social system with its peculiar lack of conception, its deep-rooted divergencies, its inborn tendency to self-criticism, its separations of economic, political and spiritual power obviously offers strong resistance to a transformation to fascist solidarity, and is aware that the deliverance which is promised would at the same time be loss of self. Thus capitalism is indeed the soil of Fascism, but the plant only grows to imposing strength if an exorbitant dose of Marxist fertilizer is added to the soil.[12]

What we have, then, is not so much a new phenomenon—left fascism has been with us for some time—but rather a new soil for a new continent on which the dry bones of inherited social theories are being nurtured and revived.

This is not to say that all distinctions between left and right were, or are, chimeric in character. Indeed, in the environment of the utopianism of the 1930s, sharp differences in attitudes between the extremes were manifest in everything from the ethnic and racial composition of the two groups to attitudes on a wide variety of measures to resolve the "social problem." As the decade wore on, pro-German versus pro-Russian beliefs became increasingly prevalent; especially insofar as these two nations embodied fundamental value orientations toward race, religion, labor, capital, and the confluence of ethical postures embodied in each nation-system. But with the Nazi–Soviet nonaggression pact of 1939, a brief period of cross-pollination and cross-fertilization came into being which reduced many older ideological distinctions to rhetoric. As a result, a set of common negative sentiments toward American democracy replaced ("displaced" might be a better term) older left–right animus. The democratic essence became the core enemy.

It was these sentiments that led to a left–fascist coalescence in the post-Vietnam climate that witnessed a nation turn back to nativist concerns. The reason for emphasizing Father Coughlin and Huey Long as "voices of protest" is to indicate that a tradition of right communism already existed; but it was nourished on infertile soil in America and, hence, was doomed to fail. The articulation of anti–big-business sentiments was often muted by more immediate concerns of social justice and emotional appeals of anti-Semitism and racism. But this was also characteristic of western European national socialism prior to World War II and of eastern European communism after that great war. The argument here presented is not reducible to some simple-minded aphorism about extremes meeting; rather, it is an effort to understand the totalitarian penumbra which contains unto itself a variety of strategies, sentiments, and slogans—all bent toward the termination of democratic societies.

Epiphenomena such as left-wing fascism or right-wing communism emerge as extreme forms of utopian visions and end as counter-utopian onslaughts. If utopianism in its essence contains an impatience with things and systems as they exist, extremist versions carry such impatience to a demand for the extirpation of evil, the rooting out of parasitical forms of social existence, a new insaturation of moral virtue in the name of racial purity, class consciousness, or historical infallibility. Thus it is that right-wing communism and left-wing fascism are identifiable with specific periods or movements. But these are not mechanistic decennial trends. They emerge as tactical efforts to deny democratic processes their due in the name of biology, history, or, simply, "the people."

What first appeared on the scene of twentieth-century history as tragedy does not reappear in our age as comedy but, quite the contrary, only as much deeper tragedy. Indeed, if totalitarianism in America remained a cultural force, in Europe it became the dominant political force.

Vladimir Lenin issued in 1920 a stunning pamphlet on left-wing communism. Several decades later, a left-wing fascism has materialized, and my concern is how, like its predecessor, it deserves to be characterized as "an infantile disorder." It was easy for Lenin to recombine elements in political society to forge new theories, yet it seems terribly painful for his followers to do likewise. For Lenin, the Bolshevik struggle was two-sided, chiefly against opportunism and social chauvinism, but also against petty bourgeois revolutionaries operating on anarchist premises. Lenin's own words in *"Left-Wing" Communism: An Infantile Disorder* are equally appropriate to the conditions of the 1980s:

> The petty bourgeois, "driven to frenzy" by the horrors of capitalism, is a social phenomenon which is characteristic of all capitalistic countries. The instability of such revolutionariness, its barrenness, its liability to become swiftly transformed into submission, apathy, fantasy, and even a frenzied infatuation with one or another bourgeois "fad"—all this is a matter of common knowledge. But a theoretical, abstract recognition of these truths does not at all free revolutionary parties from old mistakes, which always crop up at unexpected moments, in a somewhat new form, in hitherto unknown vestments or surroundings, in peculiar—more or less peculiar—circumstances.[13]

Fascism is not simply a political condition but is brought about by rooted psychological dislocations; these, linked to larger concerns, exercise an independent dynamic. The very term *"infantile disorder"* sharply focuses on the subjective qualities of fascism. Even such a politically oriented analyst as Leon Trotsky speaks, in *The Struggle Against Fascism in Germany*, of the cycle of fascism as "yearning for change . . . extreme confusion . . . exhaustion of the proletariat . . . growing confusion and indifference . . . despair . . . collective neurosis . . . readiness to believe in miracles . . . readiness for violent measures."[14] While these characteristics are invariably linked to a social class, a constant fusion-fission effect characterizes the momentum toward fascism. These terms also describe religious-political movements like Sun Myung Moon's Unification Church and political-religious movements like Lyndon LaRouche's National Caucus of Labor Committees (NCLC), the name of which was changed to the United States Labor Party (USLP).

We have so taken for granted distinctions among left, right, and center that it has become difficult to perceive new combinations of these categories. New practical political integrations disquieting the liberal imagination are hard to absorb. Lenin criticized left-wing European communism for its exaggerated emphasis on purity at the expense of victory and on vanguard putschism at the expense of mass participation, yet similar phenomena of a different ideological persuasion are taking place in America. My purpose here is to suggest the character of this recombination of political categories: how it functions in American life not simply to alter the nature of marginal politics but also to affect mainline political decision making.

While my analysis is largely confined to U.S. conditions, the state of affairs I call "left-wing fascism" is an international phenomenon. The following examples are illustrative. Massimiliano Fanchini was arrested in connection with a Bologna bombing. He first drew attention as a part of a Palestine Solidarity Committee, which he helped organize with another fascist, Franco Fredo, who was jailed for killing sixteen people in Milan in 1969. Claudio Mutti, known as the "Nazi academic" because of his spot at the University of Parma, founded the Italian–Libyan Friendship Society and helped publish speeches by the Ayatollah Khomeini in Italian. The supposedly leftist Baader–Meinhof gang, which earlier only lectured the Palestine Liberation Organization on the need for armed struggle, bought its first load of small arms from the neo-Nazi Bavarian underground. Christopher Hitchens, foreign editor of the *New Statesman*, noted that:

> There is a small and squalid area in which nihilists of left and right meet and intersect. There is a cross-fertilization, especially in Italy. Fascists often borrow demagogic leftist titles. One of the agreements facilitating this incest is a hatred of Israel; the other one is a hatred of democracy and a mutual conviction that Fascist/Communist takeover will only hasten a Fascist/Communist victory.[15]

Like all movements, left-wing fascism has a somewhat chaotic ancestry. The foremost group is what might be called the later Frankfurt school, which emphasized an uneasy mix of early Marx and late Hegel and which was most frequently, if not necessarily properly, identified with the works of Theodor Adorno. The characteristics of the Frankfurt school derive from Adorno's strong differentiation between mass culture and elite culture and his concern with the massification of society in general. For the first time, but hardly the last, in the history of Marxism, Adorno addressed a strong attack on mass culture. The obscurantist-elitist aspects of Adorno's later, post–World War II work does not refer to the democratic socialist analysis of the Frankfurt school offered by Franz Neumann, Max Horkheimer, and Herbert Marcuse, among others. Nor does it even refer to Adorno's own efforts at developing a political psychology while he was working in America on the "Authoritarian Personality" project. But to deny the anti-popular and Teutonic characteristics of Adorno's later works, worshipfully introduced in English by British and American scholars who should have known better, is to deny the obvious—and the dangerous.

Whether it be popular music or popular art, there is a clear notion in Adorno's work that mass culture is tasteless, banal, and regressive. The more remote real people were from his political dreams, the less regard did he show for the masses as such. His assumption is that such culture evolves in some abstract sense through commercialization of social classes and the existence of a worthless society. This critique is essentially antipopulist, since the emotional assault is on the masses for having such a culture. In many respects, Adorno sets the stage for a culture of left-wing fascism, which represents an attack on the popular organs of society for being what they are and a corresponding elitist demand that they be otherwise; that is, purified.

The Adorno line of reasoning, its "critical negativism," assumes that what people believe is wrong and that what they ought to believe, as designed by some narrow elite stratum of the cultural apparatus, is essentially right. With Adorno, the theory of vanguard politics is carried over into the theory of vanguard culture. The cultural apparatus is blamed for the elevation of mass culture into high culture. Attention given the so-called Frankfort school in present racial circles derives not from its origins in antitotalitarian and anti-Nazi activities but rather from its elitist outcomes: attacks and assaults on masses and their culture. The prewar Frankfurt school, with its emphasis on rationality as the basis of revolution, was corrupted in the hands of Adorno. It became a doctrine of rationality as cultural traditionalism. In this way, Adorno became central to the thinking of the *avant-garde* of left-wing fascism.

A second element of this belief characteristic of left-wing fascism is Marxism as praxis—or, without Marx, and sometimes without Lenin. This results in what might be called "praxis theory" or "action theory." It does away with the need for either an economic base to revolution, essential for Marx, or the political base of organization held essential by Lenin. What remains is a residual sentiment favoring revolutionary mysticism. The vaguely anarchical assumption is that the sum total of what one really needs is an action group, some kind of organized group of *foco*, usually clandestine, to create sufficient chaos or destruction of the state and society in selected periods of the capitalist economy. The combination of economic chaos and political protest will in itself somehow produce revolutionary action. Multinational terrorism will move into a breach, presumably magically offsetting the multinational economy.

This is often called the Cuban model of revolution, inspired by the works of Regis Debray. The transposition of a model from a small island like Cuba, with its special conditions of single-crop socialism, is quite difficult. What was originally a theory for social change in Cuba becomes enlarged into a universal theory of change; one is left with a "theory" of the vital force. The theory of putsch, the clandestine conspiratorial small group capable of seizing power at the proper moment, is common to fascism and until recently was not associated with Marxism. The theory of *foco*, or revolutionary focus, reduces Marxism first by stripping away its sense of economic forces of oppression, then its emphasis on the political sources of organization, and finally its mass base. One is left with a theory of small-group conspiracy, or terrorism in the name of Marxism, rather than a theory of Marxism as a source of social change and revolutionary action involving broad masses.

A third vital pivot is nationalism, in which the demand for revolutionary change is lodged in patriotic claims of the total system, a demand for the moment, having nothing to do with history and antecedents. Such a nationalist pivot involves not so much a doctrine of liberation as a doctrine of activity uninhibited by the need for social analysis. It is predicated on a notion of will and action at the correct moment to preserve the nation against its real or presumed enemies.

Other elements in the nationalist tradition fit into this left-wing fascist model: that every ordinary individual craves order over chaos; that one does not need a special theory of society to achieve revolutionary action; and that individual economic origins are less important than social roles. These elements from Vilfredo Pareto, Georges-Eugène Sorel, and Gaetano Mosca are incorporated into a left-wing fascist interpretation of the world, in which psychological mass contagion replaces social history as the interpretation of human events.

The sources of left-wing fascism are not abstract. Those who are enamored with appraisals of American society that seek immediate gratification and relief from ailments have become innovative in organizational form no less than ideological norms. Seeking ways to effect social change regardless of scientific or social base is the key to left-wing fascism. The effort to enlarge the cult into a state religion, the attempt to impose order and leadership on a society that seems purposively leaderless and fragmented, is characteristic of fascism—right or left, religious or secular.

In left-wing fascism we are dealing not so much with notions of traditional political involvement or traditional minor political means and ends. This involves inspiring others while servicing the needs of the actors, whether one is talking about special groups of nationalists or extreme self-styled radical groups seeking direct confrontation with other radical groups. The very act of confronting the enemy replaces any sense of organization or systems analysis. Action determines and defines one's place in the hierarchy of a political movement. In left-wing fascism the guerrilla movement replaces the clandestine *bund* as the organizational vehicle. But its impact is not simply to be dismissed because of its barricade orientation. The latter provides the basis for the militarization of politics, its decivilization, a central precondition to the fascist seizure of power.

The main political tenet of left-wing fascism is its strange denial of America and the democratic system, together with the assertion of socialism as an abstraction. Left-wing fascists have the unique capacity of being able to examine socialism without comment on the activities of Russia. They talk about the United States rather than about the loss of democracy. There is an inversion: the search for socialism becomes close to an abstract utopian ideal, but when it comes to a discussion of democracy, discourse is critically and severely linked to the United States as a nation-state. The rhetoric constantly shifts. So-called enemies are unambiguously identified as the United States and its allies. When dealing with its own allies, however, left-wing fascism turns socialism into a generalized hypothesis rather than taking the concrete forms as they are expressed throughout the world. What Adorno appropriately called the "collectivization and institutionalization of the spell" becomes the new fascist norm.[16]

Left-wing fascism accepts socialism as an abstraction but rejects socialist practice and reality and, hence, critique as a source of democratic renewal. The history of fascism in the United States mirrors that of Europe. Socialism, far from being dropped, becomes incorporated into the national dream, into a dramaturgy for redemption, for a higher civilization that will link nationhood and socialism in a move forward. This combination of words, "national" and "social," generates a new volatility. These two words together can arouse stronger and more active participation than either of the concepts taken separately.

The weakness of traditional right-wing organizations is that they asserted the primacy and value of Americanism as nationalism apart from socialist values. The weakness of traditional forms of leftism is that they have asserted socialism over and against American or national values. The potential strength of left-wing fascism, such as that practiced by the NCLC, is its ability to see how these concepts of Americanism and socialism can operate together as a mobilizing device in the development of a new fascist social order. The unique characteristic of left-wing fascism is its capacity, like its European antecedent, to combine very different ideological strains, traditional right-wing and traditional left-wing behavior, and come up with a political formula that, if it has not yet generated a mass base, has at least the potential for mass appeal.

The content of left-wing fascism is heavily based on an elitist vision of the world. At every level of society, it juxtaposes its minoritarianism against majoritarianism. It may take libertarian or authoritarian forms, but it always defends its leadership vision over any populist vision. Some examples are the hip versus the square, the gay versus the straight, the individualistic free soul versus the family-oriented slave, those who believe in the cult of direct action versus the fools who participate in the political process, those who practice nonviolence over those who assert willfulness and violence as measures of human strength and courage, those who have strong affiliations with cults and cultism over against the traditional nonbeliever (a marked departure from the anti-theological vision of most forms of leftist and socialist behavior), those who argue the case for deviance over against mainline participation in the working class or in segments of class society, those who choose underground organizations in preference to established voluntary organizations and, ultimately, those who choose some type of deracinated behavior over against class behavior and participation.

Historically, communists, like fascists, have had an uncomfortable attraction to both elitism and populism. The theory of vanguards acting in the name of the true interests of the masses presupposes a higher science of society (or in the case of fascism, a biology of society) beyond the reach of ordinary citizens. The superstructure of science, like culture generally, becomes a realm in which elites act in the name of the public. What happens to the notion of the people determining their own history in their own way? Here populism, or pseudo-populism, steps in to fuse formerly antagonistic trends. In some mysterious, inexplicable manner, these mass forces must be shaped or molded. Under communism, in sharp contrast to fascism, the stratification elements in the national culture are deemed unique or uniquely worth salvaging. But in the anti-ideological climate of the "new world," people (class) and fold (race) blend, becoming the raw materials for fashioning the new society.

Left-wing fascism does not so much overcome this dilemma of elitism and populism as seek to harness both under the rubric of a movement. Having its roots in the 1960s, left-wing fascism views the loose movement, the *foco,* the force, as expanding upon the fascist élan and the communist vanguard. It permits a theory of politics without the encumbrance of parties. It allows, even encourages, a culture of elitism and crackpot technocracy (as in the LaRouche emphasis on computer technology as a general ideology) while extolling the virtues of a presumed inarticulate mass suffering under inscrutable false consciousness.[17] The mystification and debasement of language displaces the search for clarity of expression and analysis, enabling a minuscule elite to harness the everyday discontent of ordinary living to a grand mission. Left-wing fascism becomes a theory of fault, locating the question of personal failure everywhere and always in an imperial conspiracy of wealth, power, or status.[18]

Fascism requires a focal point of hatred behind which to unify. Thus, when fascists advocate anti-Semitism, they are simply using a tactic, one not opposed by communism. It becomes a modality of affixing the climate of a post-Nazi holocaust, a post-Stalinist Gulag, and the monopoly of petroleum wealth by forces historically antagonistic to Jewish ambitions. The new left-wing fascist segments, weak within the nation, can draw great strength from "world forces" deemed favorable to their cause. The unitary character of anti-Semitism draws fascist and communist elements together in a new social climate. Anti-Semitism is the essential motor of left-wing fascism. The grand illusion of seeing communism and fascism as polarized opposites (the latter being evil with a few redeeming virtues, the former being good with a few historical blemishes) is the sort of liberal collapse that reduces analysis to nostalgia—an abiding faith in the unique mission of a communist left that has long ago lost its universal claims to a higher society. This catalog of polarities, this litany of beliefs, adds up to a life-style of left-wing fascism. Isolating any of these reified frameworks may lead to the conclusion that the dangers are less than catastrophic. But in this panoply of beliefs and practices, one finds the social sources of left-wing fascist participation and belief.

Even in the formation of the New Left in the 1960s, the roots of a left-wing fascist formulation were in evidence. Now, in a more pronounced form, what has evolved is a strong shift from a class, party, or movement concept characteristic of the sixties toward cultism.[19] We have cults not only in the strictly religious sense but in the political sense as well—marginal movements gaining small numbers of adherents but having a profound impact on the edges of society. Like the Nazi movement of the early 1920s, these left-wing fascist movements of the 1980s, such as the LaRouche organization and various socialist parties, are considered too small and inconsequential to have any impact on the body politic. But the danger to the society as a whole is that as the active element in the political process shrinks, this fringe becomes increasingly important. They do have sufficient numbers for a seizure of power once one takes into account that they do rely for victory on technology, not numbers. They rely on organization, swift movement, willfulness, and the ability to seize the critical moment. In this sense, left-wing fascist movements are not unlike the Nazi movement of the early twenties; although weak, marginal, and leaderless, they are, in fact, very much part of a social scene marked by powerful economic dislocations and putschist tendencies in segmented political processes.[20]

Once left-wing fascism is seen as an authoritarian effort to destroy the legitimacy of the established system—a series of diminutions in voting participation, party affiliation, and faith in parliamentary systems and the achievement of social goals in an honorable and honest manner—then the potential of left-wing fascism becomes manifest. This also represents a decline in traditional socialist fallback positions of mass action, mass participation, and, ultimately, mass revolution. That collapse of trust in the popular sectors corresponds to the collapse in party sectors. What might be called the bourgeois or political parliamentary pivot, on one hand, and the popular or revolutionary pivot, on the other, are both viewed by left-wing fascism as snares and a delusion, mechanisms for postponing the social revolution that is going to provide the cures of all ailments and remove all the temptations of ordinary people. Left-wing fascism ultimately represents the collapse of bourgeois and proletarian politics alike.[21] It is not only the end of ideology in the traditional sense, or an end to participation in the political process, but an end of ideology even in the socialist sense of adherence to revolutionary processes that ultimately promise organized change and social justice.

Left-wing fascism assaults both mass and class notions of legitimacy, both Jeffersonian and Leninist visions of the world, both the rational-discourse and the popular-participation models. Left-wing fascism is that unique rejection of both elements and the incorporation of nationalism from the bourgeois ideology and utopianism from the proletarian ideology. These rejections and absorptions define the four-part paradigm of left-wing fascism: for nationalism in general; for socialism in general; against parliamentarianism in particular; and against organized political parties in particular.

Left-wing fascism is much more than a political psychology. It develops direction–action frameworks in terms of specific insurgency techniques and connects them to the tradition of fascism and extreme nationalism. There is a strong element of racism and anti-Semitism in this movement, cleverly rendered by a "dialectical" pitting of Jews against blacks in archetypical terms. There is a further belief that the black movement must subordinate itself to the class structure of American life and that blacks who see their own national destinies apart from this new movement are suspect. This leitmotif of disdain toward successful blacks remains muted. The anti-Semitic modality is overt and made manifest first by its continuing fashionable currency in Russia, eastern Europe, and the Middle East.[22] Historically, fascism has had a strong anti-Semitic component: conviction of the need to liquidate the Jew as a political and economic entity and ultimately even as a biological entity. This easy glide from anti-Israeli to anti-Semitic visions has become part of the international left-wing rhetoric of our day. To move one large step further to left-wing fascism by utilizing anti-Semitism as a pivot becomes relatively simple, especially in the context of policy ambiguity concerning the legitimate claims of contending forces for national homelands.

Common wisdom has it that the most virulent forms of fascism in the twentieth century took anti-Semitic overtones. Less known but equally plain is that in light of Soviet politics from the end of World War II in 1945 until the present, anti-Semitism has been a leitmotif of the Soviet system. There is no need here to argue whether anti-Semitism is a potent force in current affairs: it is the point at which the fascist and communist hemispheres are joined. Anti-Semitism is the cement providing a crossover from right to left in terms of both ideology and personnel.

The anti-bourgeois character of both fascism and communism in an American context has now extended to an assertion of anti-Semitism as a prototype of that anti-bourgeois sentiment. Probably for the first time, Jews rather than the customary Italians and Turks are now being blamed for drug traffic, with the consequent coming together of strange bedfellows. Again, we are led back to our proto-typical organization—Lyndon LaRouche's group and its adherents. Despite the most manifest forms of racist appeal, LaRouche managed a united front with Wallace Muhammad, who, in turn, took the Black Muslims away from a black emphasis to an Islamic identification.

The effectiveness of the NCLC is seen most clearly in electoral activity and in the organization's success in building single-issue alliances with forces as diverse as the ultra–right-wing and anti-Semitic Liberty Lobby, the Black Muslims, and conservative-oriented Teamster union officials. LaRouche and Black Muslim leader Wallace Muhammad formed an "Anti-Drug Coalition" which has spread to at least eight cities. The coalition is based on LaRouche's theory that Jews are responsible for drug traffic; one happily endorsed by black ultra-nationalists. The coalition's activities include mass rallies in ghetto churches, intensive and effective lobbying for stronger narcotics laws, and seminars in inner-city high schools. Wallace Muhammad has repeatedly refused to break off his alliance with LaRouche despite appeals from Jewish organizations and responsible black leaders. The coalition has attracted an amazing range of clergymen, businessmen, mayors, law enforcement officers, state legislators, Masonic leaders, and trade union officials.[23]

This indicates the emergence of a left-wing fascism that has learned to use the techniques of right-wing fascism with impunity.[24] It has also learned to appreciate the mass character of appeals to anti-Semitism. Historically, the problem of the American left has been its narrow socioeconomic base, intellectual self-isolation, and, above all, isolation from the mainstream of workers. It perceives the working class as ready to be tapped, but only if the tactics are appropriate to the current internal situation—one in which Jews are perceived as isolated from working-class networks, gathered in the professional and middle strata of the population, and ideologically and organizationally distanced from their traditional Democratic-party moorings. Left-wing fascist elements have seen such historical circumstances as an ideal opportunity to seize a political initiative and link up with social segments of the population never before tapped.

The ambiguity of the present situation harbors the sort of pseudo-populism that can easily accommodate to fascist and social-ist ideals, both of which appeal to totalitarianism in its most ad-vanced, virulent forms. Left fascism, unlike national socialism, no longer stands for a set of rural values over against urban corruption, or for mass sentiments over against elitist manipulation. While mas-querading under populist slogans, left fascism is highly urban and elitist. The effectiveness of such marginal movements depends heavily on the state of American national interests: whether they can be sufficiently polarized to prevent concerted policy-making or sufficiently galvanized to reduce such left-wing fascist varieties of populism to manageable and nonlethal proportions.

Another serious element in left-wing fascism is its political mys-ticism, in which the cult of the group displaces individual con-science. Socialism becomes devoid of concrete practice of specific content. Socialism as negative utopianism becomes the order of the day. Real socialist practices are simply disregarded or at times pri-vately rebuked. Like satanic lodges, new groups emerge with the strong feeling that Stalinism has been an oft misunderstood phe-nomenon that deserves to be supported once again. There are now Hitler cults and Stalin cults—small bands of people convinced that history has assessed these leaders wrongly and that the source of strength of any future movement will involve a re-evaluation of these former political figures.

What I have described remains a nascent movement, an ideology and organization in the making. I am not dealing with finished ideo-logical products of large-scale political movements capable of threat-ening established structures. Nor are they necessarily a threat to classical left-wing politics. Left-wing fascism does, however, provide an answer to questions plaguing our century: In what form will fascism become attractive to Americans? What will be its national ideology? What will be its social message? Will it emerge as a wing of the established parties, or as a third party movement?

American fascism could provide a focus with a series of left-wing components: anti-bourgeois sentiments in the form of libertari-anism; fundamentalism in the form of nationalism; a defense of socialist theory with a denial of communist practice; an assertion of nationalism and Americanism as values, with a denial of mass par-ticipation and mass belief systems; elitism as vanguard populism; and a mobilization ideology in place of a mobilized population.

Most recently, the fascist tendency has been expressed as anti-government as such; a revolt against an abstraction called "Washington." These tendencies remain nascent; these fusions still remain to be crystallized in political practice. As we make attempts in the 1990s to overcome severe economic dislocations and a breakdown of organic union manifested in a hesitant attitude toward patriotism, and insofar as we exhibit a system with no apparent public commitment and only a series of specific egoistic requirements presumably underwritten by a doctrine of rugged individualism, while small groups make cynical determinations for large-scale policies and structures, to this extent can we expect to find left-wing fascism a powerful component in the political practice of the remainder of the twentieth century.

The twentieth century is polarized into diametrically opposed secular faiths. This dichotomy has taken hold because a century of war and genocide has given expression to competing messianic visions. After class annihilation comes classlessness; after racial annihilation, a triumphal master race. Subordination of the person to the collective is the common denominator. New totalitarian combinations and permutations are dangerous because they move beyond earlier hostility into a shared antagonism toward democratic processes as such. Concepts of evidence and rules of experience give way to historicism and intuition. Comfort with a world of tentative and reversible choices gives way to demands for absolute certainty. In such a climate, the emergence of left-wing fascism is presaged by a rebirth of ideological fanaticism. If the forms of totalitarianism have become simplified, so, too, has the character of the struggle to resist such trends. This awareness offers the greatest potential for democratic survival against totalitarian temptation. The United States has been singularly prone to look benignly upon the forces of left-wing fascism because it so adroitly managed to escape the real thing; namely, European fascism on one side and Soviet totalitarianism on the other. As a people, Americans are thus more ready to assume the best, not only in people but in extremist propaganda systems. As a result, the process of intellectual cauterization against such extremisms was also an overseas, "foreign," import. The migration phenomenon, with messages delivered by social scientists and historians (that is, victims of totalitarian systems) has become the essential source of resistance to the series of insidious and insipid banalities that go under the label of left-wing fascism. The work of Hannah Arendt,[25] Elie Halevy,[26] George Lichtheim,[27] Wilhelm Reich,[28] and Jacob Talmon,[29] however different these scholars were in intellectual disciplines and cultural backgrounds, spoke the identical message on one key fact: the integral nature of the totalitarian experience and the fusion of left and right sentiments in converting such sentiments into systems.

The experiences of World War II—when conflict displaced fusion and nationalist ambitions preempted totalitarian tendencies—had the effect of crystallizing differences between regimes, nations, and cultures. However, with the removal of the European wartime experience in space and time has also come the dismantling of traditional barriers between extremes of left and right. Their capacity to come together in a set of manifest hatreds for minorities, masses, and, ultimately, for democracy does not signify the end of ideological disputations characteristic of the century, but only a clearer appreciation that such distinctions among the enemies of democracy weaken the chances for the continuing survival, even slight expansion, of new lands and nations, of societies in which every individual counts as one—no more and no less.

With respect to sociology, this new situation also permits, but hardly ensures, a new possibility. Paul Hollander well captured this when he outlined three conditions that must be met for any restoration to scientific grace:

> First, sociology must regain its freedom to ask any important questions and to report findings that may offend the current politicized canons of propriety; it must be free to defy the reigning orthodoxies of the academic-intellectual community. Secondly, it must attract a wider range of people than is currently the case, including those who are inspired by genuine curiosity rather than by political-ideological agendas. Thirdly, it must embark on a new course of comparative studies (taking advantage of the new freedom for research in the former communist states) that address some of the old, but still unresolved, important questions of the discipline and the social sciences.[30]

This inspiring trio of proposals calls sociology back to its first principles. And I suspect that before we get to a new sociology, we shall have to relearn what the old sociology was all about.

In this regard, I would like to conclude with a remarkable statement by Brigitte and Peter Berger distinguishing between two kinds of ethics—one based on absolute ends and the other based on personal responsibility. They point out that

> the almost instinctive procedure of the first type of ethics is to compare any given situation with the unchanging and unerring table of values,

and then to act accordingly no matter what the consequences. Traditionalists with intelligence and a sense of compassion will, of course, repress this instinct from time to time, but it keeps asserting itself. By contrast, an "ethic of responsibility" will always be calculating, weighing intended and imaginably unintended consequences, willing to compromise with the lesser evil, even daring to be morally wrong. Mark Twain (to the best of our knowledge not a disciple of Weber) expressed this stance well: "When in doubt, do the right thing."[31]

So far have we come from such a morally centered social science that we can now hardly even cite Mark Twain, much less the Bergers, in this politically correct environment. It is tragic but true that sociology has helped tie the knot between the field and its fanatics, much less serving as a mechanism for doing the right thing.

5

From Socialism to Sociology

The origins of American sociology are bathed in religious homogeneity masked by a sense of Protestantism as pluralism. The review of the founding theorists of American sociology between 1881 and 1915 makes it perfectly plain that a certain cultural uniformity existed among figures as diverse as Lester F. Ward, William Sumner, Franklin Giddings, Charles H. Cooley, Albion Small, and Edward Ross.[1] The context of discourse rested on issues derived from evolutionism in biology and associationism in psychology. And whether the views expressed were liberal or conservative, they scarcely deviated from the norms of an America immersed in Protestantism and nationalism.

The ruling paradigm in this world of sociological uniformity was liberalism. Indeed, as Dorothy Ross pointed out in her work on the origins of American social science, it was a liberalism espousing a theory of American exceptionalism, a theory which justified an emphasis on social control as the major thrust of the discipline.[2] Such early emphasis on scientism and positivism was justified precisely because issues of control were in the hands of the social scientists themselves. The degree of consensus on the major issues of the American Century permits us better to appreciate the refusal of present-day sociologists to accept the idea of any scientific base to the field.

While American sociology always relied heavily upon European theory, such imports were made with an eye to buttressing the main centripetal forces of American society. It was a case of American innocents abroad, seeking out those concepts that would prove the worth of the exceptional features of American life. But with the emergence of totalitarianism in Europe—especially in Stalinist Russia and, more emphatically, Hitlerite Germany—the dam broke. An intellectual migration took place that changed the landscape of American sociology no less than that of American society itself. Indeed, the new intellectual migration mirrored, and gave voice to, the twentieth-century migration.

The coming to American shores of Theodore Able, Paul F. Lazarsfeld, Theodor Adorno, Franz Neumann, Marie Jahoda, to name just a few, changed the emphasis in sociology from civilizational issues to psychological issues—from "big" questions to "middle range" questions and from a sense of America as a homogeneous Protestant culture to one of America as an immigrant experience.³ But in this intellectual transformation, the inner life of sociology itself was organizationally changed. It was an imperfect transformation, one involving a level of politicization unknown to sociology's founding fathers. It is not only that the names of Marx, Weber, and Durkheim replaced those of Charles Darwin and Herbert Spencer, but also that the sense of America as a consensual experience gave way to a sense of America as a series of conflicting definitions.⁴

My purpose in this chapter is not to discuss the sociological migration of persons and ideas from Europe to the United States. This I have attempted to do elsewhere.⁵ Instead, I want to focus on a specific feature of the post-consensual, post-Protestant phase of sociology, one characterized by a concern for ethnicity in general, and for the Jewish factor in particular. In so doing, one can gain a special sense of the imperfect integration of philosophical and scientific, or, if you will, religious and secular, traditions in American sociology. Such a special angle of vision allows one to see problems of ideology in the formation of a social science.

The ambiguity of the Jewish factor in post–World War II American sociology can be seen in a sensitive essay by Seymour Martin Lipset. In the volume *Sociological Self-images*, a curious element emerges. In the American edition of this book, his essay is entitled "From Socialism to Sociology." In the English edition, which appeared a year later, the title is changed to "Socialism and Sociology."⁶ In all other respects, the article is identical. To be sure, either title is consonant with the content. But the important point is that, in an essay that emphasizes "the extent to which [Lipset's claims] have been responses to political events and to political concerns," no clear resolution is made. The Jewish commitment to socialism, bred of years of union and political struggles, did not permit a full-fledged adaptation to a sociology that stood apart from such concerns. Lipset's unresolved title problem is indicative of the more general problem of Jewishness and sociology.

The role of the Jew in American sociology is remarkably straightforward. The period of entry of Jews into this discipline follows certain events in rather close temporal proximity. Initially, it was the children of the immigrant generations, specifically those located at centers of learning within Jewish demographic concentrations, who made up the first rash of Jewish sociologists. One must add theirs was hardly a *Jewish* sociology, since the bulk and burden of the work done by this extraordinary group concerned such urgent matters as urban stratification, racial discrimination, and the specific ethnographies at work.

This remained a small penetration, at least numerically. In the interwar and war years from 1919 to 1945, sociology remained largely what it had been earlier: a strongly reform-oriented discipline embracing the vision of social change coupled with a belief in the innate goodness and perfectability of the criminal and the deviant. Without intending any disparagement, this tradition, for all its patronizing and attitudinizing toward the poor, never ceased to mean well. The field of sociology until 1945 mirrored essentially mainstream American values: mildly Protestant in religion, cautiously liberal in politics, and uncomfortably free-market oriented in economics. The New Deal, with its heavy fusion of work and welfare, firmly placed sociology in the Democratic–party camp. And given the proclivities of Jews to vote Democratic in overwhelming numbers throughout the period from 1933 to 1945, the drift to sociology became automatic—one might even say unthinking.

Given such a condition, it is little wonder that the post-immigrant generation of Jewish scholarship turned to sociology. Therein one could be comfortable among scholars with highly secularized attitudes toward science and religion, far more adventuresome when it came to economic systems and policies, and at home in a world of Marxian and socialist theorizing that looked forward in confidence to a socialism without consequences. The Bolshevik Revolution had hardly made a dent in a nativistic social-scientific tradition when it was already at the forefront of a discourse among Jews who entered the field of sociology. Indeed, the radical outpouring rampant in present-day sociology resembles nothing more than these longings of radical economics after World War I, albeit with a stronger dose of cosmopolitanism and yet a weakened sense of the specifics of the American scene.

It was less the Depression of 1929 than the European immigration between 1933 and 1939 that had the most profound impact on the parameters of sociological discourse and membership. Cosmopolitan rather than proletarian values prevailed. The German-Jewish immigration gave a stamp of approval to Jewish participation in the field and opened major centers of the discipline to them. These exiled scholars brought contextually rooted concerns to bear on big issues.

The successful end of World War II—with the liquidation of the Nazi and fascist scourges and educational grants being given to former soldiers—opened wide the doors to higher researches and removed the last remaining obstacles to Jews entering university life. Their movement into sociology, and into the other social sciences in equal measure, tore apart conventional niceties. Building upon the work of earlier generations, these Jewish scholars formed new links between local social problems and global sociological theorizing. Significant new, specialized journals reflected these links—journals dedicated to social problems, covering the sociology of everything from education to law, and even containing the beginnings of serious self-reflection on the Jewish condition.

Young scholars came from other walks of life into sociology. Daniel Bell came from journalism, Nathan Glazer from editing, and David Riesman from law; they were perhaps the most noticeable. The field soon became populated by Jews to such a degree that jokes abounded: one did not need the synagogue, the *minyan* was to be found in sociology departments; or, one did not need a sociology of Jewish life, since the two had become synonymous. This brilliant roster of postwar sociologists went on to populate departments in the great American diaspora. The University of California, Berkeley; Washington University in St. Louis; the University of Chicago; the University of Wisconsin-Madison; and countless other institutions became new centers of sociological culture.

The Jewish sociologist was, nonetheless, not quite at ease with the sociology of the Jew. At most, Judaism was a leitmotif. For example, the actual contribution of the Frankfurt School was reduced to a series of studies of the authoritarian personality; only oblique references were made to anti-Semitism in American life. But on the matter of anti-Semitism in Europe, on the special Jewish character of fascist racism, the Frankfurt vanguard remained essentially silent.

Not Judaism, or even liberalism, but revolutionary Marxism coupled with Freudianism as an ideology of liberation motivated such figures as Herbert Marcuse.

With the exception of a cluster of sociologists who were self-consciously promoting Jewish themes and values, the bulk of Jewish sociologists was much more concerned with American involvement in overseas wars, the character of racism in the American South, and the need for university reforms in general. Specifically Jewish themes were muted throughout the roaring sixties. This mainstreaming of Jewish scholarship through the vehicle of social problems made them peculiarly vulnerable. At a time when the number of Jewish sociologists is probably greater than at any other period in history, their power as a group is hardly noticeable. This absorption is the American Way, but it does stand in marked contrast to some smaller groups. Black sociologists, for example, have been promoted and have promoted themselves precisely as a unique and special vanguard force. They have come to form caucuses and rump groups which in turn have become the model for gender-related, sexual-orientation groups. Ironically, the black model has served as a stimulus to those Jewish sociologists interested in specifically ethnic themes.

At the core of the relationship of Jews and sociology is their connection to society as such. For, as is frequently the case with migrants to new lands, Jews prosper only in the presence of two conditions: political freedom and economic *laissez-faire.* In the absence of the former, as in the extreme case of Nazi Germany, a cruel fate awaits the totality of Jewish people. And without an open economic society, as in the extreme case of Communist Russia, the Jews are unable to grow as a people; they are reduced to an educational-bureaucratic mold, with little capacity to retain a sense of ethnic or religious identity. I take this as a widely disseminated datum, hardly requiring any further analysis.

The problem is that as it evolved, sociology moved away both from an open economic marketplace and from a liberal pluralism. Under the impact of Marxism–Leninism and various other conundrums and combinations, sociology shed its original *laissez-faire,* monochromatic character. But it also abandoned scientism as an official ideology of the field. The parallel growth of the welfare model—with a heavy concentration on policy from above—has likewise led to a strong swing toward socialism and social-welfare models. Thus, the shaky fate of the Jew in sociology is hardly restricted to the consequence of external opposition, such as an expanded preference for the Palestinian cause. It more nearly is a function of long-standing, unresolved, ambiguities in the relationship of ethnicity and sociology. The problem goes to the soul of a people and a character of a discipline at the same time. Perhaps we require a Werner Sombart for an explanation of the fixation of Jewish beliefs. Thus, before examining the present era, a look at the original Sombart thesis may provide a useful perspective on this vexing, troublesome relationship between a people and a profession.[7]

Rather than re-create the past, one should evaluate the present. Specifically, we should reconsider, against the background of Sombart's understanding, the global problem of Jews and their place in a variety of social systems, particularly in light of the rise and fall of modern communism as a major political and economic force of the century. But before getting directly to our topic, it is worthwhile to summarize briefly just what the Sombart thesis actually states. For like the so-called Weber thesis concerning the relationship of Protestantism and the emergence of modern capitalism, the actual writings of Sombart are more often discussed than digested.

Sombart employs a benign transvaluation of Marx's capitalism as that organization wherein regularly two distinct socioeconomic groups cooperate: the owners of the means of production and the great body of workers who possess essentially only their labor power. The pursuit of gain is the mainspring of the system; the procedure is rationalism, planning, and efficiency in the means of production. Economic activity is regulated by a cash nexus requiring exact calculations. The structure of cooperation is induced by a common appeal to the marketplace. In this context, the apparent success of the Jews under capitalism is attributed by Sombart to four factors: their dispersion over a wide area; their treatment as perennial strangers; their semi-citizenship; and their liquid wealth. But these objective circumstances are important only in the context of special Jewish characteristics. Religious values are taught to the humble and the mighty alike; their loss of a land and hope for return transformed them into a cultural brotherhood; and rationalism is the dominant ideology of Judaism as it is of capitalism. This is essentially a view shared by the best and the brightest of the age, including Max Weber.[8]

My concern is to examine this hypothesis in the context of communism, a system in which the state transcends class forces in the management and planning of the economy. The communist system fuses material and moral incentives to increase productivity and higher living standards through the elimination of presumably antagonistic social and economic forces and classes. One might argue, as do Gaetano Mosca and Vilfredo Pareto (two of the more eminent contemporaries of Sombart) that antagonisms are not so much eliminated as transferred from the economic plane of classes to the political plane of bureaucracies or interest groups. It might also be claimed that the enormous increase in centralized authority creates a ruling class out of the administrative elite. Whatever the exact empirical status communism once had in the Soviet Union or capitalism still does possess in the United States, it is clear that vast differences exist in the organization of production for the purposes of planning social goods for human ends. Beyond that, it is important to find out how the actual situation of Jews under communism helps us better understand Jews under capitalism.

In systematic terms, the issue of Jewish survival and modernity concerns Jewish conditions not so much under late capitalism as under late socialism. Living in 1993 rather than in 1903 imposes certain analytical as well as empirical responsibilities in evaluating the place of ethnicity in sociology. We must review the problem of Jews not historically in relation to the origins of modern capitalism but comparatively in light of the experience of modern communism. We need to know whether modern capitalism is as decisive a variable in the explanation of Jewish commercial propensities as has been presupposed. Better yet, we should ask whether a political *economy* of Judaism must not finally yield to a political sociology of Judaism—that is, a more broadly based consideration of state power, social stratification, and ethnic participation in both.

The question of Jews in modern capitalism must be critically examined from a comparative point of view; that is to say, how they fare with respect to Jews living under modern communism. Any serious reconsideration of Sombart's work on Jewish emancipation under economic capitalism must begin with conditions created by seventy-three years of Soviet history, forty years of eastern European communism, the experiences of Jews in Cuba since 1959 and their experiences from 1979 to 1989 in Nicaragua, and the unique position of Jews in Chile between 1970 and 1973—the Allende period and its cultural aftermath. Only in this way can one treat seriously the considerations of Weber and Sombart on the relation of religion and capitalism generally and of the Jewish people and industrialism specifically. Capitalism as a unified world system clearly was tested by communism as a world system. The former emerged triumphant; the latter collapsed. Did the ethnic forces also emerge unscathed and in triumph?

Without providing a detailed accounting readily available in the contemporary scholarly literature, it is clear that Jewry under communism moves from a state of expectation to atrophy to utter, albeit slow and painful, extinction. In the former Soviet Union the number of those identifying themselves as Jews in census data steadily decreased following the 1917 Revolution. This was a function of assimilation, immigration, or sheer silence. In eastern Europe, with the lonely but important exception of Romania, the condition of Jewish life can best be described as a series of secondary activities in the hands of the elderly and at times in the hands of curators of Jewish artifacts shown on request to visitors; in short, a continuation of anti-Semitism.[9] Cuba's thriving Jewish community in 1959 at the time of the Revolution has been reduced to a cluster numbered in the hundreds. Reports on Jewish life in Nicaragua during the Ortega period have, likewise, described an overlay of hostility for Israel as that nation which continued to supply Somoza with weapons long after most other nations ceased doing so. The three-year period of Allende rule saw a sharp polarization: "economic" Jews fled, while "political" Jews became central advisers to the socialist Chilean regime.

Theories are neither revised nor overthrown by other theories; they are modified and reexamined in the light of experience. In this regard, the overwhelming fact that could not possibly be anticipated by either Weber or Sombart was a situation in which Jews would live under an economic system defined as socialist rather than capitalist. For if Jews have certain innate phylogenetic or historical propensities that transcend those of ordinary mortals, or have certain drives that tend to underscore the normal workings of capitalism, these characterological deficiencies (or, for that matter, superiorities) should reveal themselves in socialism as well. However, the actual course of events in the Soviet Communist bloc disconfirms such atrocious notions of inferiority and superiority based on innate psychic characteristics or mystical theological properties. It therefore behooves us to examine classical doctrines of German sociology not for their generosity of spirit toward the Jews, pleasant though that may be, but in terms of how such doctrinal positions fare in the light of a century of totalitarian social systems.

Stereotypes notwithstanding, Jews have never been entirely united in their preference for a single social system, in sharp contrast to Sombart's broad assumption of a unified pro-capitalist vision within the Jewish community. Throughout German history, there was the struggle between religion and secularism; and in Russia, between Orthodox–Zionist and socialist–Bundist factions. Both indicate how limited Sombart's knowledge was of Jewish political preferences. Even after the formation of the State of Israel, such political-systematic differences remained; indeed, some might say they even sharpened. Those who believed in a strong public-sector economy and socialism characterized by the Kibbutz and the Moshav movements were in the majority, but they were opposed by those for whom Israel offered a sanctuary of free enterprise no less than the free exchange of ideas. The difference between the throbbing Jewelry Exchange on the one hand and hundreds of Kibbutzim on the other demonstrates that Israeli people and society are not exactly unified with respect to economic goals or social systems. In short, the Jews themselves historically never made a collective decision on behalf of a single economic system, neither in Israel nor in Europe. At most, they were accidental benefactors, not conscious agents, of the rise of capitalism in western Europe and not the creators of the agrarian limitations placed upon early settlements in Palestine.

In the same connection, a weakness of German Jews trying to ascend the stratification ladder is that they were unable to create any unified economic position. The Jewish proletariat was extremely active in the labor movement and in the Spartacist–Socialist movement. But there was also a Jewish bourgeoisie of considerable size and prominence—a powerful, if small, minority exercising influence throughout the period of rising industrialization. What the Jews in Europe, especially in Germany, did lack was access to state power.[10] It was the caveat of the fascists that the Jewish element lacked a sense of higher purpose or dedication to the nation. In their economic pursuits, either of proletarian communism or bourgeois instinctual behavior, they were presumed to lack a sense of transcendent national ends. In short, those who excluded Jews from ascending to power within the national context turned the facts about by claiming that Jews voluntarily surrendered participation in and commitment to a national will and destiny to which they were never allowed admittance.

By the start of the twentieth century, the Jewish Question was not particularly an economic question any longer. Both Weber and Sombart failed to understand the political character of anti-Semitism on the one hand and the national character of Jewish aspirations on the other. There is a great void in the classics of German sociological literature on the Jewish drive for political emancipation, because so much attention was focused upon the Jew as a purely economic creature or, at times, his opposite caricature: the Jew as a radicalizing political agent.

As a result, it was left to Jewish classical historians to determine the character or the degree to which specific European societies were best identified by either Jewish advancement or Jewish emancipation. In the immediate aftermath of the Bolshevik Revolution, Jews were able to advance. Between 1917 and 1929, Jews were found in disproportionately high numbers in a wide variety of political and cultural fields. Thus, Jewish support for revolutionary causes remained correspondingly high during this period. Even with the first disturbing signs of anti-Semitism under Stalinism between 1929 and 1941, the contrast of the Jewish condition in Russia to that in Hitler's Germany left little room for doubt about the marginal superiority of the Soviet socialist system.

One may infer that the economics of socialism *per se* did not hold back Jewish advances. Jews taught at the universities and defined the cultural apparatus during the first twelve to fourteen years of the Soviet experience. It was only with the post-World War II growth of Stalinism as an ideology of anti-Semitism that, by political fiat, Jewish support for Russian socialism was sharply curtailed. At that time, there began a concerted effort to prevent Jewish access to upward social mobility through education, such as it was within the framework of life in the U.S.S.R.[11] Such curtailments have also taken place throughout the history of capitalism, but in a far less systematic way. One cannot simply point to an automatic mechanical correlation between the rise of capitalism or the rise of industrialism and the successful adaptation of Jews to modernity, but it can be asserted that mechanisms of repression differ under various circumstances.

What begins to be apparent is that the rise of Stalinism, and its conscious purge of Jews from state power, was the political root of a new stage of anti-Semitism—xenophobic nationalism.[12] This was also true in Germany under Nazism, where Hitler still had a nominally capitalist system. The private sector was highly mobilized toward statist ends, but the private sector was not suppressed, simply controlled. The Jews were suppressed through this quasi-planning mechanism. This indicates the absence of any automatic correlation or fusion of a high level of market economy with a high degree of Jewish emancipation. If that were the case, it would not have been possible to anticipate the kind of negative outcomes in both communist and capitalist regimes as were witnessed either in Stalinist Russia or in Hitlerite Germany. Thus, one must turn to the political system as an explanation of and escape from the conundrum in which an *a priori* theory of the Jew is locked into economic formulas.

The tendency of classical scholars is to speak about the Jews solely as economic creatures. More recently, others have spoken of Jews solely as political creatures. Less apparent has been the rise of twentieth-century bureaucracy; specifically, the exclusion of Jews from administrative processes in Germany and their being purged from this process in Russia. A certain flowering of Jewish culture occurred during an earlier pre-bureaucratic period prior to dictatorial consolidation. However, there is no automatic correlation between a rise of socialism and a decline of Judaism or, for that matter, a rise of Judaism and a decline of socialism. To the extent that socialism, of either the national or international varieties, bureaucratizes social life, Jews are punished.

By virtue of their marginal social position, their collective morality, and their history (or whatever attribution one wishes to assign), Jews do not easily accept or comply with totalitarian or nationalist requisites. They are nervous in the face of political or ideological extremism; they are weakened under any condition in which explanations are made based on unicausal considerations of race, religion, or nation. In other words, they tend to be liberal pluralists. In any social order where collective guilt is an accepted mode of reasoning, the Jews as a persecuted group fare poorly. Thus, individualism as a fact, in contrast to collectivism as an ideology, became a Jewish mode of survival far beyond anything envisioned by Yankee pioneers or Spanish pirates.

Whether the troubles in the twentieth century come from the left or the right, from totalitarian temptations based on biological explanations under Hitlerism or historicist explanations under Stalinism, they left the Jew vulnerable. Such doctrines confer an outsider or pariah status on Jews (that is, outside the evolutionary tree or outside the railroad of history). The existence of extremism as exemplified by unlimited state power itself becomes the issue. The problem of totalitarianism, whatever its ideological derivative, is the problem to be examined.[13] The charge that Jews are rootless cosmopolitans came from all directions—from left and right—from groups to which the concepts of pluralist beliefs and voluntary associations alike were intolerable. Wherever there is a denial to multiple allegiances and a denial to voluntary associations, there is apt to follow trouble for the Jews, trouble that is aggravated by the near uniform exclusion of Jews from the administrative apparatus of a government seeking uniform allegiance.

The growth of Jewish radicalism in Europe and America was part of political integration as well as emancipation. Socialist and radical politics permitted Jewish participation in the political process outside the channels of government and administration. What distinguished German and Austrian Jews from their compatriots in England and the United States was not so much differences among Jewish ideologists but rather the permeability of Western democratic systems in England and the United States and the relative impermeability in Germany and Austria. One might argue that the United States in particular had the capacity to co-opt no less than incorporate dissenting Jewish elements into the political mainstream, but this in itself is a definition and a statement of the structure of political democracy in one context and its absence in the older European contexts. Radical and socialist politics contributed to the expansion of democracy itself in the United States and to a lesser extent in France and England. To the contrary, in Germany national socialism crushed all other forms of socialism, and in Russia a national Bolshevism crushed all other forms of socialism. But once again the essence of the Jewish Question increasingly shifts from economic to political forces as the twentieth century wears on. Thus, although the history of capitalism does not automatically bestow or enhance Jewish emancipation, Jewish existence and emancipation have contributed in many ways to the forms of capitalist development.[14] The Jewish people persevered through all kinds of social systems, but their flowering has been less a function of the economics of the marketplace than of political emancipation.

Western democratic cultures, whatever their proclivities to support or criticize the Jews, presume that belief patterns such as religion and culture are part of the private realm and, hence, protected against public or political assault. In other words, Jews flourish where a notion of a private culture standing alongside the civic culture is not only permissible but also encouraged. Where body and soul equally belong either to the state or to its religious arm (that is, a state religion), Jewish existence once again becomes problematic. The Jewish Question is not only related to the right of organization but to the voluntary character of organization. To engage in sectarian or private organizational activity of a religious, cultural, or ethnic sort involves a presumption within the dominant culture that pluralism or separation of church and state is a permissible posture.

By contrast, the unbridled totalitarian state is decisive in determining the contour and context of the Jewish Question, because such distinctions are disallowed.

Political phenomena make the Jewish Question theoretically complex, since analysis becomes multivariate on its face. Jewish life is diminished when the creative opposition of the sacred and the secular, or the church and the state, are seen as having to yield to a higher set of integrated political values. Jews suffer, their numbers decline, and immigration becomes a survival solution when the state demands integration into a national mainstream, a religious universal defined by a state religion or a near-state religion.

Jews fare well when there is no state religion. Because religious pluralism inhibits the state from taking unlimited control, Protestant cultures are healthier for Jews. It is not that Protestants are less prejudiced than Catholics. Indeed, not a few studies have established the opposite. Rather, it is simply more difficult to establish a national religion in a Protestant context. Modern Jews serve the function less of Sombart's "last Calvinists" than of born-again pluralists. They are the "last Protestant" group to gain admittance to civil society. Interestingly, in nations (even those undergoing revolutionary upheavals, like France) where such religious pluralism was negated or frustrated, Jewish community life suffered. The Jews, in turn, dramatically pluralized each Western political system. High levels of cultural fragmentation coupled with religious options are likely to find relatively benign forms of anti-Semitism coupled with a stable Jewish condition. Presumed Jewish cleverness or brilliance readily emerges under such pluralistic conditions, and such cleverness readily dissolves with equal suddenness under politically monistic or totalitarian conditions.

The Jewish Holocaust took place in Nazi Germany, a nominally capitalist economy; while the next worse fate befalling the Jews was in the Soviet Union, a nominally socialist economy. Clearly, the emergence in the postwar environment of a new national state, the State of Israel, posed new problems and new levels of analysis for older Western democratic governments as well as for the older totalitarian regimes. It also compelled social theory to move beyond a model of "rootless cosmopolitanism" in explaining Jewish survival. The actual lessons are less about Jewish survivability as a function of capitalism or Jewish survivability as a function of socialism or even growth and expansion in these terms than about the dangers of state power: the unbridled role that the modern state has performed with respect to the possible free life and association of the Jews.

The unresolved problem in the classical sociological literature is a consequence of the absence of any appropriate consideration of the bureaucratic variable. As a result, when the so-called classical positions based on Marxian premises of class conflict are invoked, they are clearly archaic. They present a simplistic approach to cultural survival by reducing such issues to psychic properties and cultural propensities. Sociology as an option to socialism nonetheless presented itself also as an option to Judaism in particular, and to a suppression of the ethnic factor generally. Sociology achieved this end by virtue of its own all-embracing demands on its practitioners.

The issues are far weightier than the turgid German sociological literature at the turn of the century described. Both empirical and theoretical parameters have shifted over time. The problem of the Jew has essentially become recognized as political in character and must be addressed unequivocally by future generations of scholars in political terms. The role of state power and bureaucratic authority provides the essential social conditioning of Jewish life, not this group's presumed innate psychic survival properties based on clever economic manipulations.

The point is not what is good for the Jews or what is good for the Germans. Zero-sum game formulas miss the point. German citizens took many religious shapes; Jews after all saw themselves as good Germans. It took a Herculean, dedicated organizational effort by the German state under Nazism to break down this association, to disengage being Jewish from being German. The role of the Jews in relation to those who controlled and managed this Nazi system became a manufactured issue precisely because Jewish integration into the German economy during the Weimar period became so complete, in contrast to Jewish alienation from the German state.

German Jews were decreasingly isolated in custom and manners as the Weimar years wore on. They began to take for granted sameness with, not difference from, their fellow Germans. Both Weber and Sombart missed this important fact. Jewish success and failure were measured in terms of German values: entrepreneurial achievement, political participation, and cultural innovation. The Nazi resentment was thus not a function of Jewish difference but of the extraordinary degree of Jewish integration into German economic business and labor. For this reason, the destruction of the Jews had to proceed lockstep with the destruction of democracy in Germany and central Europe generally.[15] Such integration with Russian life was far more restricted. Jewish power in the political process of Bolshevism was constricted by the collapse of the bourgeois and shopkeeper classes of old Russia.

Anti-Semitism became a special illustration of a rising xenophobic Russian nationalism. This fact is coupled with a belief that any form of pluralism is alien to Russian national or international ambitions. While nationalism coupled with authoritarianism leads to the dismemberment of the Jewish people as an entity within Russian life, it remains unlikely that genocide will be practiced upon its Jewish population; not necessarily because of ideological tolerance, but because there is no class or bureaucratic need to do so. The Jews were effectively disenfranchised by more than seventy years of Communist rule. They have been removed from the sources of power, military and bureaucratic as well as political. Therefore, the function of current Russian anti-Semitism, while not dissimilar to German nationalism in its claims, does not necessarily have the same dire genocidal tendencies. This may be small comfort to those Jews who began with high expectations for Russian life and the Jewish role therein. But for those who have come to measure events in terms of survival and not progress, the distinction between national communism and national socialism is important. Mistakes at this level are dangerous and not just intellectually frivolous, since the gulf between harassment and even incarceration on the one side and outright genocide on the other is as wide as the distinction between life and death.

What sociology failed to recognize was that the Soviet system did not have as its prime aim the elimination of Jewish values so much as the destruction of the democratic potential for socialism those values entailed. As a result, Jewish hopes for emancipation and national liberation were swept away as a by-product of Soviet campaigns to "de-Judaize" Russian society. The sociological curse was naïveté, a belief in the legitimacy of the symbols of socialist power. That curse reached its apex under Stalinism, which, in turn, took seriously only the actualities of state power. When ideologists of the Nazi Third Reich were locating German ills in the Jewish Question, in a quite parallel fashion the Bolshevik Right came to locate Russian ills in cosmopolitanism (read "the Jewish Question"), while the Bolshevik Left was busily arguing mythic problems of class deviation represented by anti-Semitism.

The question of Jewish survivability, Jewish growth, or Jewish inputs within a given culture can no longer be put in terms of the maturation of the larger economic system, as it was at the start of the century. It is not that economic history somehow is irrelevant, but rather that the economy itself had to be situated in a political network in which Jewish lives, faith, and thoughts come to be determined by the struggle for elementary democracy itself. We possess and have been bequeathed an economic literature but a political anomaly. Therefore, without minimizing just how brilliantly the remarks of Weber and Sombart were formulated, the dismissal of political context led to a misunderstanding of or failure to explain real-world events. That is why Sombart, in particular, fell into disrepute. His analysis did not explain the political inspiration for German anti-Semitism or Germany's blockage of Jews to the political process.

If sociology is to resurrect an analytical framework locating the Jewish Question only in the development of modern capitalism, such a framework will fail in terms of its explanatory purpose. The core issue is ethnic survivability in the development of postindustrial societies. To reify discourse in terms of modern capitalism versus a more modern communism is wasteful and a dangerous form of abstract polarization over the Jewish Question. Such a view presupposes the world of the marketplace as the only world in which Jews can possibly survive—a dangerous presupposition. Its converse, the assumption of communism triumphant, leads to a critique of the free market instead of a rejection of a bitter, intellectualized form of anti-Semitism. What is required is a reconsideration of the Jewish Question as a function of political democracy and state power. Only then will the reductionistic myth of the clever, self-serving economic Jew, a myth which helped fuel the Holocaust, finally be put to a well-deserved rest.

This brief sketch brings us to the emergence of a shocking phenomenon, shocking, that is, if one presumes the Jewish character of sociology as a given; namely, the emergence of anti-Semitism and anti-Zionism as part and parcel of the American sociological scene. To be sure, anti-Semitism remains a sideshow, a backwater of professional concerns. It is, further, an infection that attacks the weaker rather than stronger intellects in the discipline—those for whom the shibboleths of the Palestine Liberation Organization rather than the standards of the American Sociological Association are central. Yet the phenomenon of sociological anti-Semitism is increasingly evident. This phenomenon has been growing in relation to such other sociological currents as redefining criminal violence out of existence and redirecting concerns from equality among groups to special pleadings for formerly excluded and closeted minorities. Nowhere does one find such fervent espousal of alternative life-styles and the new sexual relativism as in sociology. This relativism, this ultimate repudiation of American values and norms, has a bearing on sociological anti-Semitism. The wheel has turned full circle. The Jew at the start of the century was identified as the pure marginal, the outsider, the immigrant incapable of integration. As this century draws to a close the Jew is now identified as the very apotheosis of American dominant values and culture. Jews carry the double tradition of political liberalism and economic free enterprise in a manner that Sombart would have understood, if not entirely appreciate.

Increasingly an attack upon the Jew can be registered as part of that overall critical anti-American thrust with which sociology has become increasingly identified.[16] If such sociological anti-Semites remain numerically a very small part of a stagnant discipline, their existence still served to help explain a discipline in crisis. At the very time that emphasis on methodological rigor and sophistication would have led one to expect a sociological renaissance, we have the backwaters of anti-Zionism becoming fashionable and anti-Semitism becoming tolerable.

A peculiar phenomenon happened on the way to the Jewish triumph within sociology: Because of identifying with the field so closely, the Jewish scholar has come to be seen not only as an architect of "mainstream sociology," whatever that hydra-headed beast turned out to be, but also as the apotheosis of the swashbuckling capitalist—the modernist enemy of fundamentalism. Despite the fictional character of this conceptual type, the powerful contribution of a wide range of Jewish scholars over time (from Leo Goodman in methodology and Paul Lazarsfeld in survey research to Edward A. Shils in culture and Seymour Martin Lipset in political sociology— just to mention a few) has served to draw attention to the Jew as architect of the "mainstream." Forgotten is that the critiques have also often been drawn from Jewish ranks. Such figures as Bennett Berger in culture and the late Alvin W. Gouldner in general theory, not to mention leftist figures like Immanuel Wallerstein in dependency theory and Amitai Etzioni in political sociology, have forged strong links between sociology and contemporary Marxian doctrine. But these figures in "critical" theory were, for the most part, marginal to Jewish concerns—often self-consciously so. The radical left offered only the choice of Jews without sociology or a sociology without Jews.

Whatever the actual alignment of Jewish forces within sociology, it remains the case that assaults on sociology as scientific and value-free not infrequently single out the aforementioned figures and, indeed, many others. Thus, the critique of American imperialism, reformism, and welfarism readily spills over into a critique of America's Jewish element.

The large numbers of those of Jewish background and affiliation in sociology have made them reticent to participate in the sort of interest-group politics characteristic of professional social-science life. Thus, while there does exist a talented group of people involved in Jewish studies, it has perhaps less than 10 percent of the Jewish population in sociology as such. While other minorities such as blacks, women, and gay-rights activists take a high profile, the Jewish group has opted for a low profile. In the context of increasing evidence of anti-Semitism within the profession however, such a low profile translates into silence or acquiescence. Ethnic "balance" turns out to entail putting a cap on hiring younger Jewish scholars as replacements; while "pluralism" turns out to mean uncapping every sort of vilification and vituperation in the name of hearing every viewpoint. Affirmative action translates into a subtle shift in criteria for appointments and promotions.

The phenomenon of anti-Semitism in sociology would hardly cause an intellectual stir were it not aided and abetted by sophisticated scholars of Jewish origin who still feel that placating the virulent poison of anti-Semitism will somehow mitigate its consequences. Thus, when a rather positive assessment of sociology in Israel appears in a sociological newsletter, on building sociology in a new state, it is assailed for being the work of a political pilgrim.[17] Indeed, the claim is made that Israeli responses to the Intifada were equivalent to the situation in South Africa and in Chile under Pinochet and should be likewise condemned. But beyond this are assaults on the "fascist" State of Israel, with the claim that the high participation of Israeli sociologists in the American Sociological Association is a function of "the huge U.S. aid to Israel."[18] The emergence of Israel as a nation-state, far from taming the anti-Semitic conundrums, has only intensified such attacks. The linkage of the United States and Israel has become an essential framework for combating impermeability and Zionism in the same ideological package.

Whether this anti-Zionist/anti-Jewish tendency will sprout wings and take off remains difficult to determine. Certainly, the shrill identification of Israel with American imperial interests, the singling out of aid to Israel, continued trade between Israel and South Africa—not to mention the staple arguments concerning Israeli denials to Palestinian self-determination—are all patented to elicit support from social scientists, a group already strongly inclined to support underdog causes. When this is factored into the disposition of the sociological community to liberal welfare causes, the potency of sociological anti-Semitism to silence opposition, if not gain outright adherents, remains substantial.[19]

What inhibits full-blown emergence of sociological anti-Semitism at this time is more the general malaise of the field then the specific repudiation of all forms of anti-Semitism, declining student enrollments in sociology, smaller numbers of participants in the profession, and fragmentation and specialization into more exact research pathways. But even this silver lining has a cloud: the core membership of umbrella organizations like the American Sociological Association are increasingly dominated by minority voices. Thus, a new editor of the *American Sociological Review* (William Form, who is indeed an eminent and outstanding scholar) is lauded not for his scholarship but primarily for being a "socialist-humanist." This is an overt way of saying to the sociological membership: those not inclined to profess such an ideology should keep out.

It is a matter of historical irony that a profession mired in genteel right-wing anti-Semitism at the beginning of the century should now find itself enmeshed in a far more acerbic left-oriented anti-Semitism by the end of the century. Thus, well-respected figures in sociology like Joseph Scott vigorously oppose the inclusion of Jews in minority student groups, since they are not "people of color." An additional ground for excluding Jews and for anti-Semitism as part of the new curriculum reforms is the need to include other "Semitic peoples," most particularly the Palestinian Arabs, in such offerings. This only serves to demonstrate once again the basic truths known to all: Jewish survival and growth are best served in a democratic polity and by a scientific community. What is first perceived as a struggle over the place of Israel, Zion, or Jews in world politics, and then a struggle over multiculturalism, turns out to be, at its core, a struggle over the preservation and protection of the science of sociology from the ideology of racism.

6

Scientific Access
and Political Constraints

The problem of social-science information in a Western context is hardly the lack of data. If anything, we suffer an overload of undigested information coupled with a lack of theory. Historically, the knock on European social science has been an absence of information that has permitted grand theory to go about its business unperturbed by the issues engaging the "real" world. The net result is a duality that we have alluded to earlier: positivism on the American side that sees partisan political outcomes as the "bottom line" expression of the research process; and metaphysics on the European side that manages to relegate the political process to a footnote to intellect. In the crucible of common needs and common markets, such a dualism is clearly no longer feasible.

Social science published in the East is premised on different philosophical or valuational concepts from those of the West. But even taking that into account, as methodological, if not ideological, universalism becomes apparent, gaps in the structure of social knowledge are being reduced. Historically, the "soft" social sciences, such as psychology or sociology, have been linked with "bourgeois" modalities of thought. Systematic research in such areas as demography and gerontology will be found in the literature of geography or child welfare rather than in that of sociology or political science. However, the information was, and remains, universally available, the question of quality of research notwithstanding.

Russia has abandoned earlier opposition to social research done in its own name. Former Communist authorities have come to see the importance, politically no less than scientifically, in open research procedures in these supposedly soft, bourgeois areas.

Older categories of professional disciplines, such as sociology, must yield their former status in favor of a more task-oriented agenda set not by the theorists but by the practitioners. The world of policy research is one in which background variables fade as the need for specific recommendations for change and relief accelerate. Thus, sociology, which played so vital a role in establishing public agendas, must face the music of having set in motion a tidal wave of intellectual reorientation which dissolves the discipline in favor of resolving the problem of public discourse. When reflecting on changes in policy analysis, one must start with a suspicion of policy advising and an awareness of some of the major errors or "sins" in the way policy analysts choose, define, and analyze problems. All too often policy analysts and their sponsors, the users of these analyses, react to typical issues, have a heavy investment in change, and ignore the imperatives of structural and systemic requirements.

It is clearly much easier to set priorities on who shall have access than to determine limits to access. In a democracy, the very notion of constraint raises professional hackles, because constraint involves behavioral restraints and an admission of privacy as a positive, inalienable right. Disaggregating the problem of access is the first order of business, in order to deal with specifics rather than abstractions. Similarly, reviewing the actual state of affairs with regard to access is in order rather than attempting a broad survey of moral postures toward the rights to access and to disseminate information. Doing both of these chores simultaneously will help us understand what exists now as a prelude to what ought to exist.

The right to know and the right to privacy are both embedded in the American Constitution and further detailed in the Bill of Rights. As a consequence, issues of publicity and privacy are part and parcel of a variety of mundane, or at least everyday, activities: from issues of copyright control in the entertainment industry to piracy in the publishing industry (indeed, they often turn out to be the same problems in allied industries). The access to information, in other words, is part of the cultural, no less than the legal, norm of American society. Present-day problems of government secrecy or of university science are part of the general fabric of who owns, who creates, and who disseminates information and data. The magnitudes of risk and loss change greatly under the pressures of advanced technology, whereas morals change little under these same pressures.

In the area of science policy, polarizations have occurred around the issue of who shall gain and who shall be denied access to data. The issue becomes intensified because in our environment the research and development costs for generating new products are very high, and the means to "rip off"—or for those more genteel in nature, to replicate—such hard-won data are exceedingly cheap. It may cost tens or hundreds of thousands of dollars to develop software programming for specialized use; yet it may cost pennies to replicate or simply steal these programs by illegally running out duplicate copies. The distinction between legal and illegal usage becomes more ambiguous as the line between basic and applied research itself becomes finer in a world of advanced computers and thinking robots.

What further complicates matters is that access is a geographical as well as a technological issue. The United States receives from Russia massive amounts of data through scholarly channels; that is, through scientific papers, journals, and conferences. And, of course, with even greater laxity, information manufactured and created in the United States flows into Russia in full force. In part, the issue is not scientific, because at the level of pure science, there are few, if any, "secrets." What is under wraps, therefore, is in applied areas of technology, engineering, and even administration. All of these are more nearly proprietary to the discoverer. But even here, research increasingly takes place in a "team" and multinational context. What this leads to is the mounting difficulty of a nation or a corporation to maintain product control (civilian or military) through denial of access. Restrictions exercised through the mechanisms of copyright registration and patent rights are the usual ways of maintaining secrecy. But in such areas as military technology, in which governments dominate and the marketplace barely whispers much less speaks, controls through marketplace restrictions to access are hardly effective.

There are liberal thinkers for whom science itself is endangered by arbitrary political restraints. They believe such restraints may dry up the wellsprings of creativity, thereby denying the fruits of research to broad masses of people who need information to grow and survive and creating a climate of fear and suspicion that leads to narrow parochialism and ultimately negates scientific research as an honorable activity. Arguments about national sovereignty barely dent such liberal imaginations. Instead, the universality of science is itself a major factor in establishing a peaceful global political environment, one in which sharing, rather than hoarding, of practical wisdom becomes the norm. Underneath this point of view is the belief that denial of access is nothing more than a short-term advantage, one best traded for high access and mutual respect.

For conservative thinkers, such a position represents a naive, or worse, pork-barrel vision of global competitors at the moral level.[1] The conservatives prefer to think that the gulf between East and West cannot be bridged by science and that attempts to do so only make things easier for totalitarians at the expense of the democracies. Dictatorships have a built-in advantage over democracies; they are already sealed societies, even at the perimeters. For the conservative spirit, strict reciprocity between sovereign states, rather than open access among scientists, must dominate government thinking as well as private-sector decision making.

Seen in this light, the term "access" is merely specialized rhetoric for expressing general cultural norms and values. If one believes that the Russian and American systems are roughly at parity (for better or worse), then the tendency to see open access as something good is quite high; but when one believes that such parity between the powers does not exist, then "we" and "they" modes of discourse tend to prevail. The control of access to secret or difficult-to-develop data becomes a critical focus of national science policy.[2]

It is essential to the issue of scientific access not in extreme terms—as leading inevitably either to renewed détente and universal peace or to the collapse of sovereignty and universal destruction—but in a rather low-key manner. To begin with, denial of access affects only a small portion of knowledge, information, and data. Literally thousands of professional and scientific publications travel freely between free-market and planned economies. Information travels well across borders when there is shared need for advanced development. Information travels less well on a north and south axis than on an east and west axis, despite economic and social similitudes in systems. To be sure, because many Russian-speaking scholars and scientists have a reading knowledge of English, access at the linguistic level is not much of a problem. In the other direction, major former Soviet scientific periodicals have been translated and made available to scholars in the West. Certainly, translation of former Soviet source documents and key scientific articles has been a veritable fixture in the American publishing industry since *Sputnik I.*

At the macro level of information exchange, there clearly exists a two-way street: basic scientific research and findings are communicated with relative ease. This is not to deny the existence of pre-screening mechanisms in Russia that are much more restrictive than in the United States. All research is censored by the state, and a larger portion of scientific research in potentially sensitive areas is not made available in the scholarly literature. However, it should also be noted that organizations like the Central Intelligence Agency at times also place restrictions on publication of research funded under classified auspices. Obviously, in areas of peripheral concern to national security, the amount of censorship in the former Soviet Union is minimal—except in matters of local political concern, such as anti-Semitism. (Censorship of Jewish scholarship is a common practice by Russian mathematics journals.) But this is less an issue of access than of totalitarian politics. Similarly, in the social sciences and humanities, there is virtually no censorship of documents and little problem of access.

The issue of access is complex—pitting civil libertarian concerns against military security interests. As a general rule, areas of no or little strategic military concern do not arouse thoughts about restraint or denial of access. In some areas, cooperative endeavors exist, as in medical and cancer research. But when we turn to matters of toxic gases for chemical warfare or laser technology for defensive (or offensive) antinuclear launches, then the debate over access becomes lively. However, even here a caveat is in order: there is no serious argument favoring restraint on access as a universal principle. Arguments rage over the place of sensitive research in a national security complex and the cost of such research in an international development complex. The U.S. struggle for global supremacy is many sided: with the former Soviets, it was military; with the Japanese, it is industrial. But expressed in such a manner, the issue of access is contextually located and properly focused.

The essential task is to define the problem. Otherwise, the goal of agenda setting becomes sterile, if not entirely gratuitous. We already know that the problem is not access writ large. Huge chunks of information are entirely available. We know, too, that the problem is not random access. For example, the issue is not access to a wide range of nonstrategic materials. We also know that the issue is not one of physical access. The ability to travel remains largely unimpaired for scholars—with the exception of travel to dangerous geographical areas subject to terrorist assaults in general. And here, the issue is not so much scientific access as political tranquility. The matter of access is thus quickly reduced to blueprints of strategic materials, to information that is programmable and thus readily transferable or transmitted, and to dual-use technology; that is, hardware that has commercial origins but has ready military applications.

Access seems to boil down to proprietary considerations, strategic-military concerns, and a general premise that knowledge is a commodity that can be bought and sold and is not just a natural resource to be captured in raw form. Seen in these terms, one can get beyond ideological aspects of the Cold War into serious concerns of the importance of privacy or, better, the value of invention. One can argue that in a purely Keynesian state, the price of knowledge is absorbed into the system of taxation and, hence, that knowledge should be universally granted. But neither the United States nor Russia are pure economies—therefore, the costs of acquiring access. Even if we had a harmonious global condition, restraint on access would remain constant. Issues of access are not, strictly speaking, a function of East–West tensions but of the structure of the knowledge system as such.

Thus far, we have examined the larger issues of sovereignty and economic cost in restricting access to advanced knowledge. To this must be added micro-level concerns; specifically, mechanisms of control that are involved in peer-review processes. Access is further limited by gate-keeping functions of publishing outlets. The peer-review system limits access by filtering out what constitutes poor or worthless information from good or useful information. The more exacting the scientific discipline, the wider the range of gate-keeping constraints. This is another way of saying that access is neither unlimited nor an unmitigated blessing. We neither regard every scrap of paper as sacrosanct nor demand access without a costly system of peer review.

The cutting edge of East–West differentials at the level of scientific information is peer review. The guiding metaphor behind peer review is the universality of scientific judgment as dictated by the relevant core professionals in any given discipline. It is what defines the notion of Western culture. By the same token, within the Russian network of science, although peer review operates, it does so precisely within the context of the nation. For the most part, an article or book is not considered published unless, and until, it has the sanction of an official national association. This is quite apart from whether a particular research article is considered militarily or politically sensitive. National boundaries limiting publication rather than the capacity to access information represents a distinguishing hallmark of Russian science. In this crucial regard, we can see how the very notion of "Western democratic" and "Eastern authoritarian" styles of research are fleshed out.

Thus, the Russian system of science, although sharing many characteristics with its American counterpart, reveals noteworthy exceptions: all materials published in potentially sensitive areas are subject to prepublication review by military experts, and materials in the social and behavioral sciences (not to mention humanities and literature) are similarly serviced by political commissioners, who make sure that fundamental orthodoxies are not violated. In both instances, a considerable amount of porosity exists. And in areas where closure is total, a *samizdat* (underground network) operates for disbursement of dangerous intellectual or informational materials.

In places where both Western and Eastern materials are available, price often determines what is purchased. In other words, we are dealing with a problem of information overload, and not just limits to accessing scarce data. If the advantages of an open system regulated exclusively by peer review are transparent at the advanced end of the knowledge spectrum, the advantages of the closed system are no less manifest in the mass distribution of basic literature without sensitive components.

When we look at physical access (that is, access of researchers to each other as well as to a field of investigation), a curious fact becomes apparent: East–West problems tend to diminish and North–South problems tend to expand. One might formulate this in terms of a paradigm: the "harder" the science involved, the more likely that East–West relations are at stake; whereas the "softer" the science, the more likely that North–South relations are involved. In U.S.–Third World relations, the rhetoric often turns upon phrases like "arrogance of power," "conceit of staff," and "lack of reciprocity"—the sorts of issues that are of particular concern to anthropologists and sociologists who work overseas. But in East–West relationships, one can, with equal frequency, be confronted with denied access in the name of national sovereignty, military advantage, or strategic capability. In short, two quite disparate considerations are disguised by the common issue of access to data.

We must also take into account new forms of constraint on knowledge that are a result of the rising tide of Third World terrorism; namely, denial of access to the field of research and the denial of admission to the so-called poorer or weaker nation. This is often done in the name of anti-imperialism or anticolonialism. Thus, terrorism has become a factor in the ability of field researchers in many disciplines to access their environment. It is curious but, at the same time, increasingly the case that these same Third World nation-states that demand open access to advanced technology of the West (in particular, of the United States), although arguing vehemently for closure when it comes to American field researchers or teams of investigators looking into human rights violations, argue with equal passion against any restraints on communication imposed by the U.S. government. The problem that internationalism confronts, that of isolationism, is nowhere more apparent—contradictory sets of demands notwithstanding—than in this particular Third World context. This imbalance between self-indulgence and demands upon others, although a less than politesse subject area, must be dealt with if international science as such is to survive and inform a desperate world of its manifold benefit. In other words, access is a complex problem alive on all sides of the contemporary political and economic spectrum—more so now than in the simpler times of Democritus or Alcibiades.

From my personal vantage point, the best answer to isolationism is the internationalism of science and of the educated communities. Science provides people the ability to develop a common language of discourse, not based on the particular biases within nations but rather on methodology, core research, style of work, and the capacity to place rationality itself as a goal over and above the claims of nations. This is a difficult goal to attain, because parochial and pedestrian claims have their charms as well as their forces. Scientists are able to travel all over the world, speak common languages, and read books and journals deemed universally important. Further, the scientific community has what few other communities have; namely, an institutional base from which to operate in each nation. This base is widespread enough to permit physical movement from one end of the earth to the other. Thus, the informational base and institutional superstructure permit a sense of camaraderie, belonging, and commonality to develop. This is in itself the key to forms of access. Whether the claims of the scientific community are justified or just depends in part on the claims made by other social segments, but for knowledge claims to have practical merit, an infrastructure of institutions as well as a network of ideas must be in place.

The struggle to maximize scientific autonomy is the task of scientific organizations. This is a specific mission of a scientific interest group in an environment that is terribly problematic. The administrative response to the global environment has to be addressed by political actors and the political system. A statement of policy on access to information cannot simply take the position that the needs of the American scientific community are the only needs that must be served within this total informational environment. The source of the threat does not uniquely reside in those who administer programs of big sponsored research, but also in the struggle of world empires that perform and play by different, and at times unconventional, rules.

If a scientific community does not have unreservedly shared values and norms, it cannot hope to achieve its research or theoretical objectives. However, if national commonwealth lacks the capacity for maintaining security, then scientific life may run the risk of failing to achieve a common humanity. The issue of access is a small part of this larger, terrible problem of surviving in peace and freedom. It would be fatuous to reduce a conference on access to scientific data to a demand to eliminate all varieties of secrecy. No society has survived with such unlimited access; neither diplomatic nor defense capacities would be served by such a posture. On the other hand, no society can maintain an advanced, developmental position without a free scientific environment.[3]

We are involved in a practical dilemma: one person's scientific research is another person's military security. Even setting temporal sights changes the way in which we view the value of research designs. For example, if we focus on a World War II time frame, our attitudes toward social-scientific participation is quite different than if the focus is on 1967–72, or the Vietnam war period. Basic attitudes are shaped by primary values. But if we are wise enough to put such values under an analytic microscope, we should be able to gain a reasonable perspective on time as well as spatial dimensions in our vision of social-science research.

The legitimacy of claims is made difficult because the society offers little in the way of adjudicating instruments to distinguish what is absolutely essential from what is quixotic in the way of security requirements *or* scientific requirements. In a society such as ours, both public and private-sector organizations, from the American Association for the Advancement of Science on one end to the National Security Council on the other, compete not just in the manufacture of knowledge but also in the uses of knowledge. Knowledge, more than ever, is power. But to achieve such power requires money, investment, and decisions about the needs of a citizenry. Once this broader picture is included in the study of scientific access, a modest agenda might be established, one that will still have to be fought for by the scientific community, but at least one that holds out the possibility of authentic victory. This position has neatly been summarized in an article concerning mandatory retirement of teachers. The views expressed "do not imply that intellectuals with some expertise on a subject, even if it is not scientific, cannot assist government officials by providing advice; it does, however, mean that providing advice cannot be a purely scientific activity."[4]

The dilemma of the present period is that the informal alliance between sociology and policy has turned sour. The early heady days of post–World War II, in which Daniel Lerner, Harold Lasswell, and Paul Lazarsfeld announced a social world in which there would be some sort of merger between the two in a "policy sciences" environment, proved to be neither attractive nor practical.[5] The disciplines of sociology and political science resisted amalgamation to an ancillary policy role, while the policy-oriented researchers were displeased by what must have seemed to them as a series of monkey wrenches that resulted in a derailment of desired societal changes.

The early successes of the alliance, such as were to be found in the support given by sociology in the racial desegregation decisions by the U.S. Supreme Court,[6] gave way to a series of disasters ending up with Project Camelot and the support of sociology for counterinsurgency programs in the Third World.[7] What had been forgotten by many sociologists is that contexts count for as much as contents in the manufacture of policy. The consensus behind new policies are derived less from their scientific adequacy than from wide public acceptance.

In short, when there is a public-support base for new policies—as in the case of equal rights and equal access in employment situations—policy-making becomes relatively straightforward—with or without sociological backing. But when there is a dissensus, such as in the matter of treating AIDS victims, establishing new policy guidelines becomes very difficult, whatever the consensus of professional social-scientific opinion may be.

What this serves to underline is that the relation of sociology and policy is not direct or interlocking. Only when sociology and its allies in the applied disciplines lost track of an independent position and became partisans of one or another interest group did they marginally affect policy-making; and then only as citizens not scientists. But the price of having such a marginal impact has proven very great: the forfeiture of the independent claims and status of the discipline of sociology as a sphere of knowledge.

7

Public Choice and
the Sociological Imagination

The problematic relation of sociology and policy is evident in the work of James Coleman. He is a figure of great substance and eminence who, as I shall try to show, could neither succeed in his efforts to overcome inherited dualisms and create a new synthesis based upon sociological application nor, for that matter, even earn the gratitude of his professional peers for making the attempt.[1] Indeed, for stating the obvious in statistical form—that white families flee the inner city whenever possible to escape perceived dangers and that black families follow whenever they have the means to do so—Coleman was faced with such vehement resentment and resistance that for a full decade he felt it necessary to stay out of the American Sociological Association.

That he returned to become president is a small vindication on personal terms but hardly resolved the sources of friction. Coleman has attempted to come to terms with the paradox of a professional ambivalence and a public disquiet by writing a fundamental text on the nature and meaning of sociology. In so doing, he has closed out a chapter in sociological syntheses first opened by Talcott Parsons in 1936[2] and then developed by Paul Lazarsfeld[3] two decades later. In reviewing this effort, we come to the end of the optimistic, forward-looking period in sociology and stand face-to-face with its stagnation—and possible decomposition. For empiricists could not deal with the implications of their own biases or appreciate the well of

103

resentment they set off in announcing, as did Daniel Patrick Moynihan in the 1960s, the disintegration of black family life in America.

If we deal with Coleman's *Foundations of Social Theory* as prototypical, a work of self-conscious totality by an eminent and courageous figure, we can see how difficult it was, and remains, for empirical research to cope with ideological bias. James Coleman merits attention whatever his chosen subject matter turns out to be. And when such an effort is prefixed by the word "foundations," it deserves special note. For not since Talcott Parsons's work *The Structure of Social Action* have we had such a mighty effort at theoretical reconstruction. Whatever value this effort may turn out to possess, it may be expected from a person who has spent a lifetime doing serious thinking—his own thinking. Indeed, there is a rather quaint naïveté in Coleman that I happen to share, a presupposition that honest research will triumph over every variety of distortion—from public ideology to personal neurosis. Indeed, one of the charms of Coleman is that he holds firm to this Enlightenment canon of objectivity and its rewards. As a result, the work charts the changes in a discipline, but does so with a civility and decency quite uncharacteristic of this age of sociological ideology.

One of my first recollections of Coleman was his delivering a 1968 follow-up report on the impact of busing, school desegregation, and white flight from the inner cities. As he was making his remarks before a gathering of the American Sociological Association, a banner bearing a swastika was unfurled behind the podium to signify Coleman's "nefarious" admission of what everybody in the United States, except the ideological extremists within the profession, knew to be the case: that racial differences in elementary schooling remained a social fact and inequality in educational opportunity was not dissolved by the mandated busing policy. While I do not recollect an effort to remove this hideous slander against Coleman's "politics," it did not prevent him from completing his address in a calm and compassionate way without a single acknowledgement of this disgraceful effort by a small faction of "guerrilla" and "insurgent" sociologists to inhibit free and honest speech at a professional gathering. And while this weird episode of nearly a quarter of a century ago remains peripheral to the work at hand, it must be said that Coleman continues to write with a single-minded belief that the purpose of social science is to gather facts, present them in the most adroit manner possible, and let the chips (the evidence) fall where they may. In this historical sense, the triumph ultimately belongs to

Coleman, since he is now past-president of this selfsame association.

It would have been better had Coleman's book opened with some sense of this commitment to a scientific naturalism and the limited intent of this enterprise. Instead, this important book is encumbered by exaggerated, epigonic praise, which also adorns the promotional literature issued by Harvard University Press. Coleman's own claims to "address the question of the peaceful coexistence of man and society, as two intersecting systems of action" is ambitious enough, without being adorned by puffery bordering on empty-headed flattery. While encomiums are hardly original in efforts to promote a book, in this instance they have the peculiar impact of drawing attention to those very aspects of Coleman's *Foundations of Social Theory* that are on shakiest ground.

In so doing, these fatuous paeans of praise make the task of reviewing sociological empiricism both easier and harder: easier in that one can discard the debris and settle into a consideration of the book on its merits; more difficult in that this reviewer's praise of the book for its actual worth falls so far short of the lavish remarks made in pre-publication comments that my sentiments may seem tepid and even rendered halfheartedly. Let me begin with an unequivocal statement. I urge all serious social scientists to read this book. The enormity of the enterprise coupled with the essential decency of the author earn the book serious treatment by everyone in the field. That said, what exactly is that "field"? To ask the question is to expose the central nerve endings of this new "rational choice" turn to critical examination.

To help garner an answer, it will be my aim to review the positive elements of sociological empiricism against a backdrop of claims made on its behalf. That Professor Coleman may, in fact, be an innocent victim of the exuberance of enthusiasts is a fact of authorial life. When one writes a magnum opus, one must live with claims to immortality no less than denigrations consigning a work to the historical dustbin. We should begin by clearing away some of the claims as a mechanism to permit us to get closer to this text and its author.

To start with, there is the entirely foolish and dangerous assertion in the descriptive copy that *Foundations of Social Theory* "promises to be the most important contribution to social theory since the publication of Talcott Parsons's *The Structure of Social Action*." Many extraordinary works have appeared since 1936, starting with Karl Mannheim's *Ideology and Utopia*[4] in that same year and proceeding through *An American Dilemma* by Gunnar Myrdal[5] in the 1940s, *History of Economic Analysis*[6] by Joseph A. Schumpeter in the 1950s, *Power and Privilege* by Gerhard Lenski[7] in the 1960s, *Anarchy, State and Utopia* by Robert Nozick[8] in the 1970s, and ending with *The American Political Economy* by Douglas A. Hibbs[9] in the 1980s. Rene Koenig's extraordinary volume, *Die Geschichte von Deutscher Soziologie*,[10] which appeared recently, can be thrown into the stew for good measure. While these and many other works have advanced the cause of social theory these past fifty-five years, such advancement is clearly beyond what a single text can achieve. It is perhaps time for the social sciences to abandon the search for magical formulae, the single message that will resolve and wrap up all past issues.

While it is true that an entire segment of Coleman's work covers "Structures of Action," no claim is made for a general theory of social action. Rather, the first two segments owe far more to Peter Blau[11] and exchange theory than to Talcott Parsons and general theory.[12] It is scarcely an accident that, save for a single reference to Parsons, no analysis is made of *The Structure of Social Action*. On the other hand, there are far too few references to Peter Blau's *Exchange and Power in Social Life*. This, I suspect, is a function of Coleman's "linear system of action" which takes place in a wide open social universe of interests, policies, mechanisms of control, relations of authority and trust, and, above all, a world of choice.

Preoccupied as Parsons was with European nineteenth-century systems of thought from Durkheim to Weber to Pareto, he lost sight of the voluntaristic frames in which actions take place. The boxes of Parsons's pattern variables remained hermetically sealed, not to be opened until *The General Theory of Action* in which the irrationalities and indeterminacies of Freudian psychiatry were admitted—albeit on an *ad hoc* intellectual basis and for the most part through a side door marked Paretan residues. Because Coleman makes fewer metaphysical demands upon social theory than did Parsons, Coleman is also in the more enviable position of letting in fresh breezes, not only toward a variety of theories but also a variety of practices.

A second claim made for Coleman's sociological empiricism, this time identified as the words of Jack Goldstone, is that *Foundations of Social Theory* "provides the finest solution I have seen to the vexing micro–macro problem in social science." I suspect that Coleman would like to believe these words of praise. But he is betrayed by his text itself, for it is precisely Coleman's inability to link social history with social theory that puts at grave risk any empiricist effort at unified theory.

Let us look closely at Coleman's chapter on "The Conflict between the Family and the Corporation." Since this is singled out as a highlight of the book, doing this cannot rightly be called a reviewer's selective perception. Coleman writes (and here it is necessary to quote him directly):

> In modern industrial society there have come to be two parallel organizational structures: a primordial structure based on, and derivative from, the family; and a newer structure composed of purposive corporate actors wholly independent of the family. The primordial structure consists of family, extended family, neighborhood and religious groups. The purposive structure consists of economic organizations, single-purpose voluntary associations, and governments. . . . The primordial structure is unraveling as its functions are taken away by the new corporate actors.

One could argue the empirical point by noting that family, community, and religion are not dissolving. To be sure, such micro systems are sometimes aided and abetted by corporate structures that need stability in the private life of corporate actors, no less than the loyalty of such actors in daily affairs of management. It is troubling that the private and the public are so reified in Coleman's work, that the very nuances one imagines would be made by a sociologist dedicated to the promotion of public choice are vacated. Of course, such reification is the high road of pluses and minuses, of ratio measurements in which an efficient equilibrium is imposed upon inefficient functions.

Rather than argue the empirical point, since, in fact, this is the Rousseauian observation made so brilliantly and poignantly in his *Discourses on Human Inequality*, I would draw attention to Coleman's inability to resolve the contradictions surrounding the "vexing" micro–macro problem in sociology. For if the micro is Coleman's stand-in for the individual and the macro serves that role for the corporation, then what are those mechanisms for bringing the two together once community and family are dissolved by theoretical fiat? This and a host of other dualisms are left quite intact—as fixed theoretical polarities rather than grounds for empirical explorations.

What we do not get is an attempt to understand that both family values and corporate structures are subsumed under a new variety of individualism—one in which the corporate life does not replace family life but rather both are displaced in the rush to be defined as "a person." The demands of various movements for racial and sexual equality are not just attempts at collective representations of general interests. More pointedly, they are demands that one should count as an individual—no more and certainly no less. Neither the corporation nor the family is in a position to deny this new surge of individualism, sometimes criticized, as by Christopher Lasch,[13] as a heightened egotism but just as readily viewed as an extension of constitutional demands that equity requires fair results no less than fair starting points.

Coleman does not make an effort at synthesis and does not take us beyond a rather desultory set of *ad hoc* observations. Rational choice is reduced to a woman shifting "her daytime locus of activity from the primordial structure, the family, to the purposive structure, the world of corporate actors." Here the dualism is simply left in peace. Little effort is made to situate the woman in varieties of individual self-definition; and no effort is made to understand that the postindustrial, modernist corporation is bent on accommodating or at least coming to terms with a world in which corporate identities are even less well respected and more interchangeable by far than familial identities. Where, then, is the bacon? In this case, the theory.

The formalization procedures introduced by Coleman have the effect of dampening interest in those sloppy elements in social life that fall outside the parameters of behavioral rationality. Components of social theory, such as public choice, are highly rationalized: "a set of roles that players take on . . . rules about the kind of actions that are allowable for players in each role . . . rules specifying the consequences that each player's action has for other players in the game."

The gaming analogy appears repeatedly, as in Coleman's discussion of collective behavior, the difficulties of which for "programmed strategies" are seen as interrupted by "asymmetry in an iterated prisoner's dilemma" which makes explanation higher than in a world which assumes a "tit for tat strategy." In other words, irrationalities are reduced to "panics and crazes" to be overcome by "reward structures for members of a hostile crowd." The eleven "predictions" about collective behavior are actually *propositions* about the social control of crowd behavior—without respect to the content of the demands for change or assertions of injustice. But in what sense are riots at Attica, demonstrations at Tiananamen Square, and bank panics on Wall Street similar? We are not told, because civil behavior does not descend to such baser levels of thought in Coleman's text. Rather, we are provided with models of similitudes at such high levels of abstraction that the possibilities of counterfactualization are highly improbable. Instead, just when problems become real, we are swept along into general theorems about effective social norms.

Coleman has made wide use of the public-choice models worked through in economics by James Buchanan,[14] Gary Becker,[15] and others. And I, for one, find this a refreshing coming to terms with how individual and social elements are fused in the public sector. But the rational-choice model itself has a wide variety of problems that Coleman does not address, starting with the degree of volition in *personal* choice and ending with the systemic limitations of such a model in dealing with *public* decisions. This is a shortcoming that might have been avoided, since this micro–macro management system might have been a framework for showing how sociology can augment, amplify, and correct economic utility visions of rational choice. Instead, this system is employed by Coleman to negate sociological levels of explanation in favor of an equilibrium model derived from post-Keynesian economics.

An intellectual crisis is manufactured instead of resolved. Sociology is subsumed under economics for every important area of public life. The study of self, individual, norms, authority, power, collective action, bureaucracy, and revolution yield to formal models of the dynamics of social systems. But it does so at a tremendous cost and consequence. For the dynamics are drowned in formal systems common to economic equilibrium. And the price for this formal elegance is the transformation of hard social theory into what Coleman himself describes as "The Mathematics of Social Action."

It is exactly this sort of economic linearity, this interchangeability of utility functions, this impersonal interchange system of corporate actors and collective decisions, that has led economics into its current quagmire. To push sociology in this direction, to mathematicize social functions, is entirely possible, even plausible. But the price is a level of abstraction that yields "theory" only as a formal system, not as a set of guidelines for prediction or experimentation. The hard equations thus disguise a soft approach to social theory.

As a consequence, we have a stunning outcome to this ambitious work of social theory: a series of platitudinous statements that do not so much cap sociological empiricism as exhaust what the author has to say about the subject matter of human society. The presumption is that the higher the level of abstraction, the greater the yield in predictions. But this volume, instead, becomes a source-book in the fruitless search for a social physics, with assertions at the level of the sun rising every day, rather than a sourcebook of social theory that helps explain specific national structures, specific economic systems, and specific political regimes. And in social science, if theory cannot be applied directly and organically to such matters, then its worth is obviously suspect.

Tautological theorizing is a strong thread running throughout *Foundations*. Indeed, even Coleman's conclusions repeatedly remind us that "The value of resources in a system of action is the sum of actors' interest in the resource, each interest weighted by the actors' power. . . ." Thus, if a specific issue is perceived "correctly," it becomes a problem of evaluating the public-good problems as determined by resource availabilities, and it is played out by individual and corporate actors. The problem with this construct is not the absence of such levels of behavior, but whether or not they actually exhaust or, more pertinently, really address the social life. Whether by accident or design, foundations become reductions in the hands of Coleman. And reductionism as a method becomes the stuff of social theory.

The term "macro" is little more than public-choice economics, while "micro" is little else than behaviorist psychology applied to consumer and producer behaviors. If this book were entitled *Foundations of Economic Psychology* the work would have to be described as a true synthesis of the literature, a classical restatement of the fault lines for working in the interstices of both disciplines. But the work is entitled *Foundations of Social Theory*. Hence, the reader ought to be informed as to what are the sociological, political, and cultural elements at work in such a theory. Except for a restatement of his own writings on social welfare and education, Coleman does not offer the reader any further guidelines.

We are rather told to feast, for an entire third of the manuscript, on a restatement in "The Mathematics of Social Action" that can best be described as a brilliant reworking of the propositions stated in the first two-thirds of the book in narrative form. It does not offer an empirical demonstration of the truth or worth of the propositions as such. It proves only that formalized propositions take less time and space than literary narrative. But I suspect that this is not what Coleman set out to accomplish.

This emphasis on formal procedures does work to Coleman's advantage at some levels, as when he criticizes, quite properly I believe, weaknesses in Max Weber's theories of social organization. Coleman indicates "deviations" from Weber's inability to address fully the problem of motivation in corporate life; a failure to address issues of incentive payments that modifies hierarchical arrangements; and the rise of managerialism, in which actual operations of the corporation are vested in nonowning individuals. This point, made by James Burnham in *The Managerial Revolution*[16] many years ago, does not quite come to terms with the persistence of proprietary rights in determining the life and death of corporations (see *Foundations,* 424–25).

This position, largely derived from the work of Herbert Simon and James March[17] and other organization specialists, does not eliminate annoying problems of ownership in corporate life, but it does remain the best part of Coleman's book, in that it addresses problems in classical business theory in a meaningful context. The same can be said for Coleman's summary of the literature on "how and why the state replaces the family as the major social welfare institution of society" (see *Foundations,* 579–609), although he does so by totally omitting any analysis of the voluntary or "third sector" of the economy.

The corporate structure and how it displaces the family network historically is the empirical hub of Coleman's work. Everything else is either in the nature of embellishment or simply not considered. But what if this dichotomy of individual and corporation turns out to be wrong, incomplete, or, as is already evident, too loosely stated? What then becomes of the "theory" on which this text is broadly based? Clearly, anything other than neat bivariate frameworks becomes an encumbrance to theory—and, hence, the drive toward a positivist reductionism becomes unavoidable, if not downright tempting.

Coleman is led to simplify major issues: the discussion of authority systems is framed in terms of weakening or strengthening authority, withdrawing or adding to mechanisms of control, and compliance or non-compliance in the structure of society—whether those of the State or of the Corporation. Sometimes, Coleman speaks of them in the same voice—as parallel systems of power and authority. Legitimacy, as such, becomes critical since it is at this level that individuals (micro) and collectivities (macro) link up in the fine mesh of social action.

Coleman defines the parameters of his work early on by reference to "metatheory." But upon inspection, this effort to "construct models of the macro-to-micro and micro-to-macro processes" turns out to be little more than a Benthamite utilitarianism. The great breakthrough in studying the marriage-squeeze problem turns out to be viewing marriage "as taking place in a kind of market, but one that is quite special, with each actor having only one commodity— himself or herself—to barter and with exchange rates governed by the constraint of monogamy, which prevents variations in quantity to achieve equal value in exchange." Leaving aside this rather quixotic, dare one say, profane definition of a sacral condition, this conception of the relation between the micro and the macro, or the individual and the social, is less explanatory than tautological; that is, building expected conclusions into the structure of the premises.

Coleman understands, or at least senses, the cul-de-sac in which he has been placed by sociological empiricism. As a result, the great strength of his work comes in ignoring "theory" and addressing "reality." For example, his constant drawing of attention to the absence of legitimacy as a defining element in totalitarian regimes is a breath of fresh air. But it derives not so much from Coleman's vision of exchange of actors or demands for effective norms as from a democratic stand against nonrational authority and against authoritarian systems as such and, perhaps, against constraining theories. Therein lies the great strength and lasting value of Coleman's book. He is sound enough in his moral judgments to rise above his own sociological explanations.

The huge size of this volume notwithstanding, *Foundations* is, in the final analysis, a simplistic account of social life. There are few people in the text, only "actors"; few authoritarians, only "authority systems"; few actual nations, only massive undefined "states and societies"; and few corporations, rather "corporate structures." When examples are provided, the text comes alive. One delights in comparisons of the Philippines with Afghanistan, not because of any great profundities, but because reference to such nations serves to illustrate points sharply and without the heavy boot of formalization procedures.

Where sociological empiricism addresses situations that arise from specific milieus, it is strongest; as in Coleman's treatment of the electoral process as a mediation between individual and social choices. Thinking about primaries as transforming a single-stage process of choosing among multiple alternatives into a multistage process of paired choices is a good way of describing nuances in the democratic process. The further explication of how voting preferences determine not only outcomes but also sequences of events is likewise an elaboration on how democratic processes work in advanced societies (see *Foundations*, 414–15). But of course, all this is expressed at a level of social abstraction that virtually ignores Democratic and Republican parties or, for that matter, liberal and conservative belief systems.

The analysis of student rebellions in terms of cost-benefit analysis, the estimate of probabilities of victory in the act, the role of the nonparticipant, how revolutionary movements develop alternative systems of authority and not just a divestiture of authority as such—all of this, extends the Hobbesian world by showing how revocation of authority is a struggle for the establishment of new lines of authority, a search for legitimacy that is as true of student revolts as of national revolutions (see *Foundations*, 496–502).

The most interesting and impressive elements of sociological empiricism are related to the conduct of social research as such. In candor, it must be noted that these chapters have little if anything to do with the essential theses of the book, but much to do with Coleman's professional and personal battles to instill social science as the ground for realistic policy-making. And here Coleman is on solid ground as a serious worker in the vineyards of sociology. But this hardly constitutes a mandate for general theory.

When policy is elevated to a formal construct, sociological empiricism becomes problematic. Coleman examines the place of sociology in a study of how welfare programs operate, of how the American soldier helped determine changing patterns of national integration. He examines why those in authority use social research more than those who hold power. His answer, "in a single word—legitimation" (see *Foundations*, 640), is fair and worthwhile but hardly begins to exhaust the nuances of the social system. Repeatedly, Coleman lets opportunities to explain the relationship between general theory and concrete application slip by.

Sociological empiricism properly argues the case that social research is important when it has an effect on social function, and that even if this runs a risk of bias, it is worth pursuing. One wishes he would take the argument further and make the case for reflexive mechanisms that adjust for and overcome biases in the research design and not just a claim for the rational person pursuing rational models in a rational environment. Throughout the text, one feels the presence of an urbane and civil scholar, a principled person who stubbornly pursues the quest for knowledge as a way to inform social policy. But the relationship between the morality of such a social science and the theory of such a social science remains terribly obscure. One feels a recourse to a Smithian hidden hand underlies much of the effort at synthesis. The pursuit of self-interest as rational choice somehow makes possible social interests as a rational system. That all the inherited problems of *laissez-faire* economics thus come to the surface seems not to be disturbing to Coleman.

Coleman does state frankly the need to move from sociology to a "new social science"—one that addresses problems of a demand for knowledge irrespective of fields of training and helps to realize opportunities in the transformation of society, applied research, and social theory to form a linkage that becomes "a new social science, appropriate to the new social structure" (see *Foundations*, 663–64). However, the substance, the specific fault lines of a Baconian *novum organon*, a new social science to go along with the new social theory, is left to the imagination of the reader. It is as if Coleman himself is reticent to draw in bold relief the implications of his own exercise.

When we get to the synthesis of new social theory, sociological empiricism becomes a restatement of first principles; only not so much in grammar and narrative as in the mathematics of social action. The last third of the book becomes a relentless restatement, in formal and elegant terms and far more condensed as one might imagine, of what the previous two-thirds announce. The synthesis, the new theory, the new social science, becomes an amalgam of public goods, public choices, organizational frameworks, and individual actors intersecting with corporate goals. While this may represent a triumph of Kenneth J. Arrow[18] in economics and R. J. Herrnstein[19] in psychology, I see only a return to the problems that plagued economics and psychology at the start of the century and stimulated the growth of sociology and political science. Coleman's coda and classification of older theory is a synthesis of what the civilized person as *homo Americanus* would choose. However, public choice is not a resolution of what ails the social sciences today.

Calling critical attention to the weight of the formal over the historical may appear to be an unfair criticism of a work aiming to provide a general theory that can serve many occasions, peoples, and systems. But I think not. After all, even such a master of microsociology as George C. Homans[20] could infuse his every page with rich historical examples and a worldliness that, frankly, Coleman simply does not evidence in this volume. Bringing people back into social research is precisely what Coleman fails to do. To be sure, it is sociological rationalism at its weakest point, at the point of subjecting reality to the test of theory instead of theory to the test of reality, that permits Coleman to come closest to the individual to whom this book will be repeatedly compared—whatever the propriety of the comparison.

One is left with the distinct impression that sociological empiricism in its empty battles with sociological rationalism, economics, and psychology emerges as the victor; sociology and political science, as the vanquished, or at least the vanished. The aura of reductionism is reinforced by the absence of examples from non-Western cultures. It might well be that such an outcome is warranted on the basis of the evidence before us. But this does not rest on theory so much as informational retrieval and reliability. It is not enough to ask us to fall back solely upon the strengths of economics and psychology in the shared life of the human race. In the absence of an articulated set of empirical reasons for why this decision should be made, Coleman's case remains unexplained and, in part, inexplicable.

Nothing less than human culture as such is dissolved in formal sociological subsystems of corporate life: actors receiving benefits—marginal and total, acting and not acting, contributors and noncontributors. If social theory can be formalized in such terms, then the book is a success. But if much of social life escapes the net of social theory as civic behavior, if the sum total of generalizable experiences of various societies is more than can be processed by such a consensual model, then the book must be measured as less than satisfactory.

As measured by the criteria outlined by sociological empiricism, Coleman's work is a successful venture into the organizational known. My own preference would have been for less formal elegance and for many more forays into the vast disorganized unknown we call society. The latter alone yield the sort of anomalies and paradoxes that permit the production of new theories and make possible new synthesis. As it stands, *Foundations of Social Theory* tells us much about the state of affairs in the research world of economists and psychologists but precious little about the state of affairs in the messy world of ordinary men and women. It was said of George Bernard Shaw that he was a good socialist fallen among Fabians. One would have to say of James Coleman that he is a good sociologist fallen among economists. He tried to drown out the noise of his detractors by remaining true, after a Bolshevik fashion, to the actual state of affairs in U.S. society. Instead, he only served to draw attention to the very inequities he strove mightily to overcome.

What emerged in Coleman's *Foundations* is less a synthesis than an analytic schema sure to be picked up by others and converted into yet another new fashion. Sad to say, it is neither entirely convincing nor quite novel, yet it has managed to capture a large share of a shrinking sociological marketplace. But the inability of so major a figure to come to terms with sociology as a source of ideology no less than inspiration tells us much about the sorry state of a profession. For when push turned to shove, the sociological empiricist perspective, the reliance upon evidence, models, and just plain data, proved inadequate. Empirical sociology as such was no match for the onslaught of true believers who invaded the field—not as a process of science, but rather as a source of radical destruction presumably to be followed by revolutionary construction. Sociological empiricism was done in by myopia, turning a blind eye to the fanaticisms within by making frightened appeals to the evidence without. The final obstacle to the decomposition of sociology was removed—by the very "politics" orthodox sociology in America had so cleverly come to scorn. What began as an effort to construct a sociology without politics, ended in a politics without sociology.

Ultimately, the problem with the rational-choice model is that philosophically it is predicated on conservative premises that the individual is sovereign and that decisions made by the person are inevitably superior and carry greater beneficial effects than those made by collectivities or states. However, sociologically, it is based on theories of fiscal utility—that is, given a chance to opt for that which will provide the greatest good for the greatest number and, at the same time, satisfy personal proclivities, that person will arrive at a rational choice which is also a public value.

The difficulty is that this approach marks no significant advance beyond the classical eighteenth-century model of Adam Smith and Claude Helvetius or the early nineteenth-century efforts of Jeremy Bentham and James Mill. It may bring the "hidden hand" into the open, but it is still not clear that actual decisions in the public square are made in this manner. Moreover, establishing a theory of public values denies a theory of political interests. And in the clash of interests and values, it is by no means self-evident what constitutes a rational decision or, for that matter, whether a common ground uniformly exists. Recently, several rational-choice theorists have sought to "make a better job justifying" their theory;[21] but the problem, as Randall Collins comes close to admitting,[22] is not the justification but the theoretical model as such.

As a result, this sort of sociology banishes itself. It cannot accept the prospects for basic conflicts that lead different people, structures, and agencies to radically different conclusions. It cannot accept as a goal for sociology the examination of paradox or, even more important, that the very essence of the discipline is to make as few *a priori* judgments as possible as to what constitutes rationality or irrationality and to stay close to the study of behavior as social construct.

There is another, more pedestrian dilemma: the public-choice model has led this group of sociologists (and an impressive number of economists and political scientists as well) to form their own journal and their own frame of intellectual reference. *Rationality and Society* boasts of being "the new frontier"; nothing less than the very "kernel of explanatory theory in social science." But in so doing, Coleman and his cohorts, replicate and enlarge the Balkanization process of the field as such. They become one more sectarian group of several hundred individuals, involved in the process of self-validation and self-praise. The Rational Choice group joins the Sex and Gender group, the Marxist Sociologists, the Political Economy of World Systems group, and countless others in developing a specialized niche and, hence, has little chance to enlarge the public discourse it is so interested in serving.

To its credit, the Rational Choice group is fully aware that sociology is in crisis, that new parameters of theory and research must be established if the field is to survive, much less expand. But its members remain ambivalent in seeking to establish common ground with those in the other social and behavioral sciences. Thus, at the very point of replacing social structure with public choice, this group repeats the Parsonian error of seeking a unique pedestal on which to place the sociology they dismantled! Pretentions to being interdisciplinary notwithstanding, the professionals from economics, political science, cognitive psychology, and game theory are described as "scholars from other fields." Coleman's *Foundations* is still trapped by the imperial pretensions of a sociological field with a rich past but an extremely risky future.

8

Social Contexts and Cultural Canons

Every era redefines what of the past remains relevant and what needs to be discarded. This is hardly news, but it is not exactly an immutable process. Individuals who are discarded in one period are sometimes rediscovered in the next; the finite amount of information that one can carry about dictates as much. Civilizations, like individuals, travel with only as much intellectual baggage as they can manage, and perhaps a trifle more than can be comfortably handled during such transitional periods as the *fin de siècle* of the twentieth century. The current reconsideration of what from the past has meaning today, then, is hardly anything new. It forms the core of what Mannheim referred to as the sociology of the culture.[1]

What is new and special is the hyper-consciousness with which this cultural redefinition is being conducted. The redefinition of the fault lines of culture is not taking place between the institutes of higher learning and the larger nonacademic world but rather within the bowels of university life as such. It is this tension within that makes the struggle over the canon so relentless, so bitter, and so dangerous. Indeed, the most prominent feature of postmodern academic life is an awareness of the transitory nature of what we once deemed permanent and even the unauthentic character of what we once considered perfect. Everything that has transpired is subject to criticism, then to revision, and finally to renovation. If this process is untempered by the rather vague rules of civility, canons of the past simply become "canon-fodder" of the present. And this is as true as much for those who seek to bury Marx because of the debacle of communism as for those who earlier sought to ignore Thomas Aquinas because feudalism had been superseded by other social orders.

The tradition of the new is linked to criteria and standards by which society, or a community, decides what to discard and what to add to the durable repertoire of ideas. Multiculturalism, and the search for a new canon, is fundamentally about criteria of selection and mechanisms of control over society as a whole. The argument centers around who is to be represented rather than what sorts of achievements are to be recognized. Multiculturalists do this in a frontal assault on presently accepted standards of assessment of achievement, arguing that these standards cannot be understood apart from principles of domination and power.[2] The struggle over definitions itself becomes a struggle for power.

Polarized expressions of self-interest as well as self-consciousness are the hallmark of our times. It is little wonder that the debate over multiculturalism and the new canon is taking place in a sharpened climate and is articulated in life-and-death rather than in genteel have-or-have-not terms. Whether or not we are better off as a result of this triumph of coercion over civility, the need remains for some sort of analysis of why this ferocious polarizing process occurs.

What role, if any, is to be played by social science as a canon unto itself—a set of beliefs conditioned by cultural limits rather than a set of experimental facts? For now, I should like to defer this question in favor of the larger issues as they have been framed. For even if we agree that the future of social science is secure (no small presumption to start with), the question of which parts of this social-science tradition are living and which are dead or, as is more fashionable to say these days, which parts are authentic and which are fraudulent merits urgent attention.

The current debate over the canon of Western culture must be distinguished from those ongoing changes that are precipitated by comments and criticisms within a given cultural or scientific paradigm. In the past, this debate was characterized by external attacks rather than internal critiques. By external attacks, I mean those assaults on intellect and its carriers that come from pressure groups, agencies of change, and even institutes with interests that may or may not correspond to the world of learning or the canon of culture. The absorption of such assaults within university life is new and different.

Behind the myriad discussions on Eurocentrism versus Afro-centrism, or what is termed monoculturalism versus multicultural-ism, and the various other platitudes ending in "ism" is a debate over economy no less than an ideology. Decisions about what to teach, and on what tracks, are, in effect, sly determinations as to who should do the teaching. Indeed, the common rhetoric refers to the conflict over the canon as a struggle for property rights, as in who controls knowledge as property controls the minds of the young. It is, therefore, hardly an accident that this stormy rhetoric about canons and cultures is taking place within the groves of academe—the one institution in Western culture where market forces do not entirely hold sway.

The cry against dead white males and the assault on Western culture provide a rationale for demanding new approaches toward academic hiring and promotion that can uniquely favor those who advocate Afrocentric culture. If Afrocentrism is an entity on a par with Eurocentrism, the next step may well be parity in hiring. Aca-demic institutions have tacitly accommodated to hiring quotas. This debate is thus in large part an effort to provide an ideological legitimation for pushing harder for minority hiring in academic life—but with the added caveat of doing an end run around tradi-tional qualifications of talent and performance. In this way, ascrip-tion comes to displace achievement as the essential academic mea-suring rod.

This is not to claim that the consequences do not have profound trickle-down impact. A recent New York State report would support massive curriculum changes in the public schools, affecting every family with children in the system. At stake in this debate is every-thing from whether Columbus was a great explorer or a mean-spirited imperialist to whether Thanksgiving is a day for American reflection and celebration or one to bemoan the fate of the Indian tribes. These discussions are most intense in urban schools, where racial tensions are highest; but they have permeated the entire school system.[3]

A further glimpse of the economic undertow to the debate may be gained from the bitter struggle over the government's decision, in April 1991, to defer federal reauthorization of the Middle States As-sociation of Colleges and Schools on the ground that such accredita-tion demands "diversity standards." This means that evaluations are based on affirmative-action hiring programs and the multicultural content of the curriculums. The issue has been posed politically in terms of the independence and civil rights of the accrediting boards on one side and their presumed encouragement of race-based hiring quotas at educational institutions on the other. But few doubt that the basic concerns are who has the power to hire, to legitimate hiring practices, and, of course, what criteria are to be used in filling academic positions.

That the schools singled out for Middle States Association's denial of reaccreditation are places like Baruch College (a division of the City University of New York that emphasizes business and traditionally has had a high Jewish student enrollment, a place that can hardly be called a conservative institution in any politically acceptable definition of that term) is by no means beside the point. The impact was immediate. Two black professors from Baruch lobbied the Middle States Association's evaluators to censure the college; charging the inevitable "institutional racism," followed by a dismissal of "white" definitions of "excellence," and concluding with a characterization of university scholars as "pseudo-scientists" who foist white domination on the students. The second professor warned the university that it was jeopardizing its reaccreditation. Almost immediately thereafter, the chancellor of the City University elevated a black dean to a vice presidency. Affirmative action plans were expanded. The president of Baruch, Joel Segall, resigned, and a black-activist president was chosen in his place. The crisis was quickly resolved and Baruch's accreditation was renewed.

The issues involved in the struggle between the Middle States Association's accreditors and the U.S. Secretary of Education have been well covered in the press and have been cast in terms of the limitations of an external agency in determining hiring policy and curriculum definition. But it is clear that behind this florid rhetoric of multiculturalism are two decades of affirmative-action programs in which political agendas displace academic quality.

Hardly a university or college in the United States is not engaged in head counting. Academic appointments for select minorities are often stalled on the basis of scholarship but forced by administrations in terms of community needs or ascriptive criteria. Many institutions have leveled percentages of faculty (but not of minority student enrollment, it is worth adding) at slightly below the percentage of these minorities of the larger population. But this has, if anything, sharpened, not abated, discussions about standards in academic hiring. At the same time, other groups traditionally underrepresented in academic life have made large inroads. Despite continuing wage differentials, the advancement of women, in academic life as in business, has been significant. One might speculate that women's

achievements are a factor in these demands for new criteria for dis-
tributing the "goods" (jobs) in society.

By making the issue multiculturalism instead of opportunity
based on achievement, these ideologists have contrived a new mechanism whereby the place of women can be subsumed into an assault
upon the dominant position of white Europeans generally. Of
course, few radical feminists will participate in assaults on the curricula of "dead white males." But some women may see themselves
as vulnerable in this assault on the canon and its white European
representatives and perceive it as a threat to their new-found power
in the academy.[4] In short, instead of a settlement of the spoils in a
more equitable way, the new conditions of spiritual struggle over the
canon seem to intensify the struggle over such material goods between a black minority, which sees itself as still holding the short
end of the stick, and all other claimants to the spoils.

Another traditionally underrepresented group, with high levels
of achievement, is Asians. Their case is awkward for those committed to multiculturalism. Asians have not railed against "Eurocentric" culture. Japan and Korea clearly want to be considered ex-
tensions of the West, while honoring the historical achievements of
their civilizations. A case could be made for the neglect of teaching
Asian cultures in the American curriculum. But Asians have not
turned this into an issue; they clearly believe that twentieth-century
culture, for better or for worse, is Western culture. Many vanguard
administrative groups have dealt with Asian achievement by fiat;
that is, by statistically ruling them out. Since increases in minority
faculty over the past twelve years have largely been of Asians, the
faculty data are now reconfigured in such a way as to eliminate
Asians from hiring considerations. This issue already surfaced at the
University of California during the current decade.[5]

A primary device for breaking the hold of achievement-oriented
criteria for hiring is to attack the very notion of the American idea!
For if this is a nation with a unified culture (albeit of distinctive
national, ethnic, racial, and sexual modes of presenting that culture),
the question of who teaches becomes subsumed under larger valuational concerns. However, in a world in which America is not larger
than the sum of its parts, but is only its parts, Afrocentrism is readily
juxtaposed against Eurocentrism. The intellectual grounds for employment and ultimately hiring policies become double-tracked.
The racial features of the teacher rather than the subject matter
taught can be made central. The struggle over maximizing the Afrocentric course offerings can then be seen as a struggle for black

professors—unless one is prepared to believe that multiculturalist advocates will accept white teachers as fit to teach Afrocentric subject matter.

In the world at large, that is to say, in the world in which cultural artifacts like records are bought, books are sold, and films are rented, the question of the canon is readily resolved. Each piece of art or invention is unique, but its ultimate value is largely determined by the contribution it makes to the larger picture. For the next world of social science, the study of social class or of political authority is a path well trodden. But how such themes are explicated in specific studies of unique conditions determines the worth of the canon as well as of the study. This is not to deny the emergence of fault lines that change the social-science canon. For example, Simone de Beauvoir's book *The Second Sex* did just that. It moved the study of gender to a new plane. It also cleverly undermined the idea that class alone determines social positioning in advanced society.[6] As a result, this text broadened the base of analytic tools considered appropriate in the study of society as a whole. The canonical text is thus an organizing premise, or often a reorganizing premise. Why and when such activities take place has many explanations. That measurable fault lines in culture exist and change is unmistakable.

A canonical social-science work may release a flood tide of derivative works of uneven importance. It may also lead to demands for changes in the larger society's system of rewards and punishments. But this is a creative process from within, not a mandated process from without. Were it the latter, those with power would steadily and unfailingly command the outposts of science and culture. Clearly, the sociological rhetoric of oppression notwithstanding, this is not the case in open societies.

In a field like music, where canonical fault lines are even less problematic, one may identify as a "classical" context the works of Bach, Hayden, Mozart, Beethoven, and Schubert. But if the context switches to twentieth-century classical music, then the canon is narrowed to Stravinsky, Shostakovich, and Prokofiev. If the context is narrowed further to national idioms, then for the United States, for example, one would identify Copland, Gershwin, and Ives as central to the musical canon. Values may be universal, but contexts are specific and particular. For that reason, one cannot mandate a particular art object or musical work by one group or nation upon another group or nation. How culture travels is a subject of study unto itself in the history of ideas. That there are limits to such travels is simply a social fact.

If we switch from global characteristics to idiomatic ones, say from classical to jazz forms, then Louis Armstrong, Jelly Roll Morton, Art Tatum, and Charlie Parker would be dominant figures. If we move further into a selection between individual soloists and large groups, then Duke Ellington, Benny Goodman, and Count Basie become the essential canon. And obviously, if we define the canon in terms of the instrument, say the jazz guitar, then people like Wes Montgomery, Kenny Burrell, and Django Reinhardt are critical. There is no reason for confining this to music. But since this is an area in which black performers and creators are heavily and brilliantly represented, it indicates that argument about canons is only partially about tradition and continuity in culture.

The social-science canon has to do with a measure of shared-domain assumptions. To speak of a canon in social science is not the same as a laundry list of literary "greats." Rather, it is to speak of a cluster of people within each profession who form the nucleus, the core, about which fundamental discussions on the nature of the discipline and what it seeks to describe are built. The social-science canon, thus, has to do with a measure of shared-domain assumptions. It is the ability to discourse upon the world through a shared intellectual framework of those who have defined a special field of inquiry—wherever they may live or work and whatever may be the racial character or the religious profession of the actors involved.

If one speaks of contemporary macroeconomics in the United States, it is appropriate to take for granted that the speaker—be it a male or female, a black or white, a young student or an old professor—is familiar with the writings of Robert Heilbroner, Murray Weidenbaum, Seymour Melman, Kenneth Boulding, or Paul Samuelson. Indeed, it would be hard to imagine any discussion of the economics of, say, the costs and benefits of an arms race that did not include the thoughts of these people. In that sense, they are part of a contemporary (and perhaps a short-lived) canon. Whether these or others will stand the test of time depends not on ascriptive features—on whether they are Jewish or Gentile, for example—but on how well their ideas continue to frame the thoughts about the costs and benefits of military preparedness.

The same is true in other areas. In the study of contemporary American society, it is presumed that one knows the work of Talcott Parsons, of Robert K. Merton, and of C. Wright Mills. There is no presumption that all will walk away from these works with a unified vision of the nature of power or of the relative importance of science, religion, and technology in the structure of a social system. But there is a presumption that to be a sociologist signifies a basic knowledge of these men and the information as well as theories undergirding their discourse. In that sense, one presupposes a canonical capacity of all concerned to make discussions of social systems and social structures more meaningful.

The canon changes as the context of discovery shifts. If one is interested in the dynamics of small-group behavior, in how the psychology of the individual and the sociology of the institution meet in direct contact, then the works of George C. Homans, of Freed Bales, and of Herbert A. Simon become central. For such contextually driven concerns are only minimally dependent upon what may be presumed to be canonical in other parts of the sociological forest. To be sure, there may be a great struggle over the propriety of one set of domain assumptions over another. In such a way, the struggle over canons may escalate into a conflict over basic world views. But this is part of the life of science and culture. When it is mandated by external interest groups, the potential for dogma and doctrine grows exponentially.

Can one seriously imagine the study of sexuality without reference to Sigmund Freud, Alfred Adler, and Carl G. Jung? Or put another way, can they be dismissed because they were from Europe or were men? Clearly, whether in Berlin or in Buenos Aires, the conduct of psychiatrists and psychoanalysts rests on their achievements. Further, without them, there would be no Helene Deutsch or Anna Freud or Marie Bonaparte; that is, particular inquiry into the nature of female sexuality. The denial of a canon in this field is tantamount to a rejection of a mentoring situation that clearly prevailed and reached its fruition in a field like psychotherapy, in which self-analysis and the process of learning are so powerfully intertwined.

The same holds for subdisciplines like political theory. One presumes that a literate person in the field has a working knowledge of the works of Hannah Arendt, Harold Lasswell, and Leo Strauss, among others. For how else, and in what other terms, is one to discourse on the nature of authority, the character of political persons, or the connection of moral purpose and national interest. It is conceivable, perhaps imperative, if the focus is international law versus national power, that one would substitute Hans Morgenthau, Hans Kohn, and Hans Kelsen for the aforementioned trio. What becomes inconceivable is that such figures could be dismissed from consideration because they were German Jews. Yet it is the disquieting trend that the break-up of older canons derives from newer configurations of power; not greater sources of wisdom.

In special areas of research, such as the study of the black condition in America, where the canonical works are those that have been seminal, obviously, in such a context, the works of E. Franklin Frazier, W. E. B. Du Bois, Booker T. Washington, and John Hope Franklin are critical. Whether such works can widen the context of their impact is a function of their level of universality in meaning and applicability to other areas of research. To reduce these issues to fratricidal struggles between black and white, male and female, Jew and Gentile vulgarizes and denigrates culture and science to forms of expression unto themselves. It is to convert the struggle for a democratic culture into a footnote to contemporary arguments about who gets what and why. John Hope Franklin[7] has referred to this process as the reduction of scholarship to folklore.

The oft-registered complaint, widely reported in the press and echoed at the base level, that we learn only about European and white, male Americans and do not have African Americans and Asians in our curriculums, fixates the issue of the canon on ascribed status. It further removes the question of learning and respect for standards from the intellectual and academic agenda. For if a person or even a discipline can be dismissed on the basis of national origin, then the very prospects for a universal culture, a shared set of standards and norms, is likewise denigrated and ultimately destroyed.

To be sure, there is a problem, one that distinguishes culture from science. It is easier to argue that, black or white, the response to the tragedy of *Hamlet* will be common to all who share in the human experience of ambiguity in action. Can one argue that the same commonalty or universality can be established for *The Protestant Ethic?* It is fair to say that the answer is a qualified yes: all people can be moved psychologically by the plight of a young prince, but perhaps the English-speaking reader, or for that matter the Danish student, would be slightly more moved. The same is true for the study of religious conditions under which market economies emerge. This issue, too, might strike the European Protestant with particular force. But by the same token, Catholics, Jews, and Moslems can well appreciate how this distinctly European phenomenon affects their own societies. Indeed, the new pluralism is itself a source for the breakdown of sociology as a universal language and shared discourse.

One can multiply fields to fit the canons or reverse the process by multiplying canons with which one is familiar and hence to understand fields of study and realms of discourse. But to take refuge in the silly idea that a canon is a law of nature and then deny that social scientists speak about laws of nature quite misses the point of it all. For the canon serves to frame discourse by sharing identities. It liberates the imagination. The brain is not locked into worship of the past, but selectively takes from that past what is required in the present. To deny such a vision is self-defeating of the idea of social science.

The worst that has happened with the canon is the sterile idea of the "greats," of a group of people cast in stone, who stand heads and shoulders above their compatriots and who still guide our destinies in a godlike way. This is a secular form of the lives of the saints. The busts of past figures of learning—some important still, others barely read—were romantically aped in the earlier part of this century and even more throughout the nineteenth century. Scholars were transformed into saints. An assault on this kind of canon is healthy and salutary. The need for heroism may be genuine, but the assumption that this need is an outcome of hero worship is unfounded—it merely represents a substitution of hagiography for biography.

In short, the world of culture offers not a single canon, but a variety of canonical frameworks. Once we address actual fields and figures and move away from thick abstractions like "Eurocentric" and "Afrocentric," it becomes increasingly apparent that serious discourse does indeed reveal the existence of major and minor works in all fields of endeavor. Each of these provide common layers of meaning to discourse and experience no less than intellectual relationships a class of people have with one another. Democracy means a choice of canons, not just a hierarchy of great works on one side and a cacophony of noise on the other. Multiculturalism argues against the idea of a monolithic canon and for recognition of greater and lesser, broader and narrower, canons.

The assault on the canon is a subtle assertion that the academic world and the ideas it promulgates are some kind of hustle to keep privilege intact and the disenfranchised out of the system. The insinuation of ascribed characteristics of race, gender, and ethnicity serves to emphasize the trade-union character of university life, but in doing so, the notion of such a life as special or deserving of unusual considerations—such as three-month summer vacations—is put at grave risk. While the age of medieval castles of learning may be at an end, this attractive, but simple-minded, assertion of equal pay for equal rank and of identifying world-class scholars among blacks, Puerto Ricans, other Hispanics, and American Indians must be understood as precisely the reverse: a denial of even the existence of the notion of world-class scholars or that it can exist in a world in which special interests and sheltered categories call the academic shots.

For those believing in laws of nature in science or objective facts in social science or choices in ethical theory, the options of a world without canons is unthinkable. This would be to reduce the world of knowledge to a vicious nihilism in which the anarchy triumphs in spirit and the behemoth rules in state. Unfortunately, such nihilism quickly transforms itself into a new dogmatism, into a new canon as it were. In such a world, Stalinist genetics and Hitlerite anthropology prevail. The behemoth once again triumphs. In this sense, our choice is what the canon should be, not whether we will have one. The argument is about the criteria for selection. Demands for multiculturalism sometimes disguise appeals to nihilism. It is one thing to argue the need for broadening the cultural base of the canon, it is quite another to argue that any notion of a canon is itself spurious. In the rhetoric of the moment, one hears discordant notes on these choices.

The attack on the canon, whatever the canon involved turns out to be, is, in point of fact, an attack on the idea of learning. Such political assaults claim that truth is relative, subjective, and partisan. If science is reduced to a chaotic heap of unconnected and contradictory assertions put forth by a special class of stratified white males, then science itself becomes at best a matter of majority vote and at worst a matter of raw force. Those in opposition to the canon clearly prefer the path of raw force—since even a recourse to numbers might cost them a victory. The problem with such a narrow self-interested approach is that if this is a choice in favor of backwardness and convention over modernity and innovation, we will all be losers.

The thing at stake is, ultimately, the culture, not the canon, or better, the canons. The risk is the substitution of one work for another in one course or another. But can science and culture survive with litmus tests of political, class, ethnic, racial, or sexual loyalties? The forces that would mandate from outside what ought to be taught are the same that have now led to the decomposition of fields of learning in the name of ideology. The saving grace is that in a democracy, people always have a way of defying authorities and traditions and of selecting those works of the past and present that represent and image the world in which we all live.

Each age must rediscover the dialectic of culture: a canon is built upon a contradiction. Rousseau understood this and noted in his *Discourses on Inequality* that the price of civilization is nothing less than a diminution of individual freedom. Marx understood this and wrote in *The Communist Manifesto* that the destruction of the morals and manners of medievalism were made possible only by a cruel industrial system in which labor displaced faith as the touchstone of life itself. And Joseph Conrad in *Heart of Darkness* wrote the same at the turn of the century, noting the terrible human costs of bringing Western civilization to Africa through the imposition on African natives of an imperial system more brutal still than had been the factory system for European workers.

Each in his own "canonical" way, Rousseau, Marx, and Conrad well understood the high price of progress in a civilization and culture. Indeed, they each worked mightily to reduce the terrible and painful costs of such ubiquitous changes but never denied their necessity. It is because such individuals from the past understood the necessity of pain in the process of gain and the equal requirement of a constant struggle to eliminate or at least reduce such pain that they are exemplars of modernity. The canon of canons, if one dare use such a phrase, is to appreciate the multiple sources of values, not to pontificate in the name of dogma which value is correct. For ultimately the cultural canon is not a thing but an awareness of a process. The ideal of the good, imperfectly represented in the realities of a morally relative universe, will yield only to a better canon. The sheer weight of blustering force, or the power of bully pulpits, will scarcely touch the signposts by means of which an open society determines civic life, political culture, and, not least, the social sciences.

Part II

THE RECONSTRUCTION OF SOCIAL SCIENCE

If a new idea system does appear to give new life and impetus to the realities of contemporary Western society . . . it will be the consequence of intellectual processes which the scientist shares wih the artist: iconic imagination, aggressve intuition, each given discipline by reason and root by reality.

Robert A. Nisbet*

* From *The Sociological Tradition.* New York: Basic Books, 1966.

9

Reconstructing the Social Sciences

Like many people in social science, I have given an inordinate amount of time to examining the fates and fortunes of our corner of knowledge. Professional narcissism is one of the vices of self-reflection. Thinking about one's profession can occupy a disproportionate share of attention with respect to its actual worth. This caveat notwithstanding, the current status of social science is of increasing concern not only to sociologists but to some outside the field as well. I would like to outline a few central problems; a few areas of contention; and beyond that some new aproaches for improving the current state of affairs.

Let me begin with the presumed statistical fact that areas like sociology and political science are on a long-term stagnation. Indeed, if one factors in the erosion of active full-time members and their displacement by marginal increases in student members and foreign affiliates, then the word "decline" would be more appropriate. Consider political science, taking as a baseline the twenty-year time frame between 1973 and 1992. The number of regular individual memberships in the American Political Science Association in 1973 stood at 12,250. By the end of 1992, twenty years later, it stood at 12,000. This is hardly a promising trend in an association that has embraced the belief that progression and retrogression are numerically determined.

If we examine the statistical information supplied by the American Sociological Association, an even more profound pattern of membership devolution is exhibited. The figures show the same kind of tendency that is evident in political science. Sociologists are apparently modest about supplying numbers (or their meaning). The total membership in 1980 stood at 12,868. By the end of 1992, the membership roll listed 12,300. Again, a pattern of long-term stagnation and decline is evident—one that could not be covered by increasing student and overseas memberships. Indeed, full members declined from 9,333 in 1980 to 8,604 in 1992. It is evident that sociology is in a period of permanent stagnation; attempts to put a pleasant spin on this matter notwithstanding.

Comparisons between political science and sociology in terms of declining membership patterns are noteworthy, since the "softer" disciplines have suffered disproportionate loss. This information is readily available, there being nothing magical or secretive about organizational membership. There are those who have drawn premature and erroneous conclusions from this data; the most superficial is that there is a general crisis in the social sciences. But the burden of my remarks states essentially the reverse. I do not believe that there is a general crisis, but rather some specific crises in select areas. Information has been drawn in too partial a manner. We have to look at the broader picture to see what is living and what is dying in the social-science communities, numerically and statistically to be sure, but also intellectually and ideologically.

I am prepared to argue that the social sciences are not collapsing, only changing. The so-called crisis in *social science* as a whole is largely spurious. Growth areas in everything from area studies to urban studies abound; and these areas involve enormous numbers of social-science participants. The crisis in sociology (and, to a lesser extent, in political science) that does exist has resulted because a great deal of social-science activity, even the majority of excitement and interest, is not only in new fields, but in areas that once drew strength and inspiration from sociology.

The social-research environment is becoming much more specialized and quite autonomous from older centers of learning. As a result, an enormous amount of fragmentation has occurred. Historically, progress in science shows its face first in fragmentation and only later in integration.

Let me begin by noting a sample of organizations which have either emerged or grown rapidly in the last decade, the period of time during which the precipitous decline in the number of association members in sociology and political science has taken place. They are, in random order, Social Science History Association; Society for the Scientific Study of Religion; Association for Public Policy Analysis and Management; Inter-University Seminar on Armed Forces and Society; National Association for the Education of Young Children; American Society for Public Administration; Society for the Psychological Study of Social Issues; Society for Applied Anthropology; Society for International Development; the Evaluation Research Association; the American Society for Information Science. This list can be multiplied ten-fold. There are at least one hundred new organizational forms that have emerged during this period of supposed social-science decline. This should caution analysts and doomsayers about the methodological fallacy of drawing too narrow a sample.

If the only organizations sampled are the American Sociological Association and the American Political Science Association, the limited data will inevitably lead to inferential errors about the status of social research as such. What we need, then, is a more ample notion of the universe inhabited by social science. One has to examine the questions of innovation at one level and the problem of continuity at another level. We must begin by asking the question: where does change take place within the social-science community? Increasingly, changes are taking place in highly specialized environments. Information-networking, computer-utilization environments have especially strong task-oriented goals in such interdisciplinary efforts as crisis management, public-policy research, and decision-making processes. Each of these are research frames exhibiting high crossovers and reintegration.

Problems of new dam construction in the Netherlands are characteristic of the changing role of social science vis-à-vis the new technological environment. The Netherlands, by virtue of a special geography, has its own activities outlined in a unique way. One-third of the Netherlands has been reclaimed from the sea through the enormous will and imagination of its people. They kept exploring ocean shorelines instead of plundering neighbors. But the further out they went, the harder it was to reclaim land; the process requiring greater work and imagination. Decision-making now involves the utilization of land planners, geographers, technologists, engineers, urbanologists, all getting together to solve the problem of reclaiming the ocean floor for human habitation.

Let me draw your attention to another, quite different area of the professional world. Having had twenty-four operations for plastic surgery, I have always been curious about how plastic surgery as a profession has evolved. Thus, I receive literature issued by the

American Society for Plastic Reconstruction. The first couple of pages of one recent brochure were astonishing. When I was being operated upon as a child in the 1930s, there was only a surgeon assisted by a resident medic. They did the repair, or the reconstruction, and I was returned to the ward and then went home a few weeks, or months, later. But this brochure described how every plastic-surgery operation now involves a social psychologist, a social-welfare expert, and even estheticians. Surgeons are now part of a team, and the operation is now a means to an end—human reconstruction—rather than an end in itself.

These two disparate and random illustrations show us that the environment of professional life in social science is different from what it once was. The social-science world has shifted to meet obligations that either were not present or were not recognizd twenty years ago. What I am talking about is not only an organizational proliferation for the sake of multiplication, but also an entirely new involvement for social scientists in all kinds of things that formerly we were either not involved with or involved with on an ancillary or marginal way.

One might argue that such newer areas of research and application are not really part of social science. But I think such an appraisal would be dangerous and shortsighted. Optimally, the purpose of operations research or evaluation research has an entirely social-scientific purpose: to measure and monitor human activity, accurately and fearlessly. Not all professional life is infused with idealism, and so, for example, not every physician in the world operates within the framework of the Hippocratic oath. On the other hand, for a group of professional people not to have that optimizing scenario available—no measure of what they can do and how well they are doing it—would be sad indeed. While the concerns of measurement are to minimize the amount of cheating, lying, and anything else that goes on in that environment, a field cannot be defined by its cheats or its liars. In every area, in every subdiscipline, there are going to be minimal and optimal levels of performance. Thus, the extremes of error-free or error-prone parameters cannot define a field of work, at least not in the social and behavioral sciences.

Recent trends clearly indicate that the output of doctoral programs in sociology exceeds the demand for college and university teachers. As a rule of thumb, for the next twelve to fifteen years we should expect only about two-fifths to one-half of new Ph.D.'s to enter teaching. Thereafter, as college enrollment turns upward, the academic demand may rise, but in the immediate future, graduate programs should adjust their curricula to prepare up to one-half of their graduates for careers in nonacademic settings—research, counseling, administration, and other forms of public service.[1]

Targets for research careers include urban planning and development research, demographic rsearch, operations research, marketing surveys, and research for public and for-profit agencies on a wide range of societal problems; for example, drug abuse, crime and delinquency, marital adjustment, child abuse and child development, mental health issues, aspects of demographic change (including hunger and regional development), and a host of others. In addition to substantive courses, the methodology of research, computer programming, data management, information sources and data evaluation, survey methodology, operations research, mathematics and statistics, and interviewing techniques should all be made available to graduate students in social science. The training for interdisciplinry and team research should not be overlooked. The problem shifts from too much or too little theory to the types of theorizing needed to accommodate this new work environment.

We are at the point now at which over 30 percent of sociology's personnel are not in academic life. This is an enormous increase from a much lower percentage ten years ago and from hardly any thirty years ago. The growth of social science, then, is not something that takes place only within the academy but basically goes on outside the university. It is an anomaly, perhaps a tragedy, that what we do in the university oftentimes is unknown to non-academic practitioners who either do not realize what is going on in university life or do not seem to care.

There is no use looking at old nineteenth-century social-science categories: psychology, sociology, political science, anthropology, and economics. Today we have to examine social-science innovation in quite a different way. We have gone beyond an Aristotelian framework into the Copernican world.[2] Aristotle was that extraordinary figure who developed the taxonomy within which science lived for two thousand years. But that taxonomy in relation to ourselves is no longer feasible. The organizations in these five areas dissolved early in the century. They no longer encompassed and they no longer defined the character of the field. For example, the eclectic American Planning Institute giving way to a multiplicity of area-studies organizations conveys a sense of the diversification in which its members are involved. In this process of combination and recombination, some social sciences will have the capacity to adjust, others will not.

Professional personnel relate to one another in terms of real interest; that is to say, if interest shifts from Soviet studies to Latin American affairs, your primary focus of identification will no longer be with the American Political Science Association but, perhaps, with the Latin American Studies Asociation—not for everyone, clearly, but for many. What we are dealing with, again, is new organizational forms, the erosion of old organizational forms, and new organizational tasks. If you view it in this way, what we really have is a veritable renaissance of social science. The word "renaissance" may appear extraordinary hyperbole at a time of considerable pessimism, but if we enlarge the scope of what we do, I think it is clear that the word "renaissance" is pefectly applicable. The numbers involved in the social sciences tell the story. The number of people has increased from less than seventy thousand in the major disciplines in 1950 to close to four hundred thousand. So what we have is a critical mass of people who can be called social scientists doing a variety of tasks in a wide number of areas. This is a cause for a certain amount of optimism. What we also have to face is the kind of social science this may produce; that is to say, what kind of social science is stimulated when dealing with large numbers and a vision of the world that is oftentimes not defined by the social-science community itself.

Our history gives us a strong impulse to define our own problems. But increasingly we face a British situation in which the tasks are defined for us; that is to say, since we are dealing more and more with the concept of service, the question of whom we are servicing increasingly comes to define that area. In this peculiar environment, a kind of positivism prevails. An enormous amount of social-science work is produced by a large number of people doing a great variety of things, but there is also a declining amount of theoretical input into that framework.

At the methodological level there has been a great deal of forward movement in social science. One can utilize a textbook in psychological statistics with good effect in sociology, political science, or economics. There would be some relabeling, and the language will change slightly depending upon the particular field of reference, but essentially we have reached, albeit slowly and painfully, some common language in terms of the way we massage and handle information. At basic levels one could take a methodology course in psychology and have it meet a sociology requirement without any difficulty (or vice versa). The issue is joined at the metatheoretical level: what does one do with information, beyond its mere collection and organization?

One problem is that of departmentalism. Bureaucratic administration is a very difficult issue to work through. There are many ways in which it is confronted, some of them are very harsh; for example, does one permit a department to shrink numerically until it realistically reflects outside concerns and interests? Likewise, popular areas can be expanded to meet new demands. In a market economy, multiple demands may obscure the significant from the trivial. There are dangers in that. The marketplace expresses mass opinion and policy dynamics, but not necessarily scientific truth.

The marketplace, nonetheless, does get what it wants. The real problem is can the university be supple enough to change and meet the requirements without becoming overwhelmed by a kind of crude empiricism. It cannot avoid confronting the fact that there is slippage in the older disciplines, nor presuppose that all that is required is to join the forces of darkness; that is, all one has to do is become part of the marketplace. This seems to me equally dangerous and equally erroneous. This is a difficult, transitional moment in social-science history.

In order to retain a sense of general purpose, we need general learning. Consequently, this means that a need for a university environment that offsets commercial trends and policy-targeted work is greater than ever. There will be points of contention. There are, however, important reasons to maintain this medieval institution called the university for the purpose not only of doing pure research untainted by partisan agencies, but also of providing watchdog functions with respect to the quality and the range of services performed in the marketplace by the social sciences. I would have to add, in all fairness, that this idealistic view of the university presupposes the existence of nobility in the university that is harder to locate than to think about.

The university is the last large-scale institution to be brought into the market economy. The academy has been the only agency, up to a relatively recent time, that was able to survive outside of a cost-benefit framework and bottom-line mentality. But now, no university president is unaware of cost factors in running his institution. Everyone is concerned with faculty productivity. We all remember when such a phrase would stick in our nostrils. The very thought of productivity violates the terms of the old academic ethos, but now it is a commonplace—number of hours spent with students, in classrooms, in grading papers, factoring in summertime vacation in estimating value of academic work, factoring in winter vacations and estimating whether one is on a ten-month line or a twelve-month line.

Every faculty member has been converted into an economist. Under these circumstances, the gap between the university and a so-called "real world" has shrunk. Being apart from the real world has certain value; medievalism is a good thing; being able to waste time is an important condition of creativity. The entire Aristotelian theory of science is based on the notion of leisure time—without leisure there is no science. If the university becomes a place in which everybody's time is measured, including sleeping time, then the continued existence of science itself may be in question.

Having said this, the question of the university's value itself changes. Our concern is not whether to adjust to the needs of the larger community. This question, oddly enough, has to be turned on its head: can we stave off that community at some level; are we able to inform that external environment which supports us (that is, the taxpayers) that academic researchers are worth the cost of waste? It is an extremely hard sell. It is difficult to tell a teamster who is working across country—twelve hours a day, twenty hours some days, one-week vacation, maybe two—that a university professor is somehow entitled to special consideration and concern. One might make a case for this differential, but it will have to be done in terms of performance. In saying this, I return to the other end of our problem; namely, productivity. University faculty, university administrators, and university students have an awareness of this. I can recollect a time when the happiest day for students was a day when they did not have to attend class. But today's students are almost as aware of the bottom line as administrators. They expect the professor to be there if they are going to be there. This is a different environment from what academics have experienced in the past. At some level this solves a problem by not solving the problem. Oftentimes, we seek policy mandates when we do not require any. The world has conspired to solve this problem for us.

Great Britain is a very helpful and illustrative model. The British have not had a profoundly important general theorist since Herbert Spencer. Yet the bang that the British get for their social-science buck is much greater than the Americans get for theirs. Richard Titmus has noted that "for very little money and with informal party affiliation, British social scientists can go all the way to the top with recommendations." These are often acted upon.[3] However,

British social research often operates in a theoretical vacuum. Not since Herbert Spencer have the British had a profoundly important general theorist, and hence there is a lack of any theoretical design. But avoiding a Parsonian model-building may also have proven to be a blessing in disguise. The relationship of social science to social policy provides a serious problem, one that potentially has the capability of dividing the house of social science. The increasing tendency toward positivism is hard to stop since that is indeed where funding is, where the support base is, where the activity is going on. There are those people who claim, perhaps with equal concern, that we offer no larger sense of a vision of society, a vision of the world, a modeling structure; nor possibilities of revising the system. Positivism may solve middle-range problems but leaves intact large-scale considerations of the system itself.

How do you get out of the positivistic bubble? If social scientists are operating within this sphere, they will modify it, sometimes enlarging it, sometimes shrinking it, always reorienting it. But will they ever see anything beyond the bubble? The question is answered, in part, by the emergence of a counterrevolutionary trend to this positivism which has been very strong among a highly vocal minority. Basically, the issue at one level is whether you can get beyond the policy research environment: can you presume at some point the policy does not work, and if not, what will? Are those kinds of concerns outside the confines of the bubble itself, or are they in the research environment itself? This is an extremely important problem that the social-science community must address, and indeed is beginning to address.

There is a fine line between innovating and imitating; the former represents overcoming the drag of tradition without entailing surrender to an external market environment. It is difficult to balance an inner history and an inner logic of a discipline and meeting the tasks of the larger society at the same time. There is no magical formula; but there is at least a possibility of recognizing the thin line between positivism and idealism in social science. This is a delicate surgical procedure; if we fail at either end, that is, if we fail to meet the challenge of modernity, we will fall apart and be without students, resources, or labor power. If, however, we meet the challenges by simply replicating what everyone else is doing in terms of evaluation or policy research, then we run the risk of losing the game of social science itself, losing the ability to perceive long-range phenomena outside the bubble within which we operate. Thus, there are two ways of losing big, and only one way of winning small.

Another dilemma is what might be called the problem of rationalization. Social scientists are often employed by government agencies, no matter who is in power, Democratic or Republican administrations. The rationalizing impulse of social science is to maximize activities that are economical, that cost less and produce more, that produce better labor output, and that produce higher levels of productivity. Economists, indeed, are the quintessential rationalizers among social scientists. Positivism maximizes rationality of the system, reaching maximal levels of equilibrium wherever and whenever possible. Any good economist must have that as the essential definition of his activity. However, sociologists' (the "softer" types, social-welfare-type people) concerns are somewhat different—their architectonic, their dynamic, is somewhat different. They want to maximize welfare or personal happiness or equity concerns among sexes, among races, and among people, even if it costs more. From a rational economist's point of view, the Environmental Protection Agency is a very expensive hobby. You have to believe that the environment is worth the high cost of preserving it in order to have that kind of input.

The conflict is between those for whom social science is an activity to rationalize systems, make them function more effectively, more efficiently, and those for whom social science is maximization of benefits. Ezra Mishan's classic work on cost-benefit analysis, has fifty-seven chapters (he is a good economist, by the way); fifty-four chapters are on costs and three chapters are on benefits. His whole vision of the world of social science is reducing benefits, maximizing activities, and making costs go down for services rendered.[4] I cannot imagine a sociologist writing a cost-benefit analysis with the same kind of proportionate disequilibrium. So, within the framework of the social sciences, we have different architectonics at work. The growth of the activity indicative of a genuine burst of enthusiasm and excitement does not necessarily mean that we are going to have a unified social science.

Does this signify that social scientists are some new breed of debased conservatives who are protecting the status quo, or are they in fact men and women of such virtue that they must guard their purity against worldliness? That is exactly the problem that Weber dealt with when he examined the notion of *Beruf* (that is, the notion of "calling"). What exactly is our *Beruf*, what exactly is our calling?[5] Let me turn the issue of calling around. Instead of asking questions about the ethics of sociology, we must ask about the sociological basis of ethics: what do we know about the evolution of ethical theory and ethical judgment in the context of social analysis? In what contexts do people make certain kinds of ethical decisions rather than others? If we ask the question in this way, if we ask the question in terms of the import of ethical decision making and how people arrive at their ethical frameworks, we will be in a much better position to understand the issues of ethics and construction, destruction, and positivism and criticism.

In the last ten or fifteen years, every social-science organization has "gotten religion." All at once, social-science associations have developed extensive ethics codes and boards to monitor such codes. Some are so long and so complicated that they make the American Constitution look like a simple document. They have an extremely complex kind of ethical framework within which they presumably operate. Were there no ethics before this? Is the ethical codification a consequence of growing moral concern, or a function of professionalization and exclusivity? Do the ethical codes work? Do they promote good social science or frustrate it? These are perfectly reasonable questions for social science to ask about the character of ethics codes. In that way, social science has a legitimate entry into the issue of ethical doctrine as such. If we ask the question: "what are your ethics?" all hell breaks loose. Ten people in a room invite ten different moral persuasions. Everybody gets angry at everyone else. This strikes me as a fruitless way to go about a discussion of ethics. We have to be able to develop a frame of reference where social science adds its own insights to ethical analysis, and we have not yet achieved that goal.

The question of ethics has become a big item on the national agenda since Watergate and Vietnam. Yet there are very few texts on the subject of law and ethics. What is the relationship between a legal profession and ethical codification? I would review the rise of this profession in relation to a set of personal obligations and needs. But that is exactly what is needed, and that is the way I would do it.

Consider *Brown vs. Board of Education*, in which social scientists not only predominated but prevailed—you would not have *Brown vs. Board* without the social-science input, without the Kenneth Clarks, without Gunnar Myrdal. This is a small victory for progress but one in which the question of law, social science, and ethics can at least be focused upon in a very precise, exact, and meaningful manner. The literature is there, and it is a significant literature.

The question of reversing the decline of enrollment in sociology departments is another kettle of fish. The decline of enrollment in older departments is permanent unless those older departments become attuned to many of the newer issues and technologies. Enrollment decline can be curbed only when departmental decision making begins to reflect the newer tendencies taking place in the social-science community. When evaluation research is attached to a certain department, sociology, economics, or psychology, may act as an umbrella for newer trends and tendencies. When they function well, older departments are administrative devices for incorporating newer tendencies and blending them with older, more established modes of thought. Innovation is the way in which a profession survives and grows. It is one thing to shout that the world is not the way we want it to be; quite another to make jobs for people who are perfectly decent, who are your own students, and for whom you feel a keen responsibility. We each share those larger problems, but we each also have smaller, move intimate requirements.

We share an environment where mass bases and mass democratic participation are limited and restricted. There is a great deal of talk about mass democracy, but it is more mass media than it is mass participation. We live in an environment where decision making is becoming increasingly more intricate, complicated, and dangerous. The larger the society becomes, the fewer are the number of people involved in decision making. If social scientists deal with the larger environment as a kind of Americanized version of neo-Platonism (for want of a better term), we are faced with some serious issues. The danger is that we have eroded participation and any capacity to influence a society. Our society seems uniquely responsive to the scientific and to the social scientific. Scientists have become the magicians of the twentieth century; the shamans of the era. By the same token, science becomes subject to scorn and riducle when the "magic" does not work.

There are all kinds of tendencies and trends in government, but there is no doubt that as a collectivity, social scientists gain wide access into places barred to ordinary citizens. There is a role for placard carrying. One need not argue against mass participation to state the case for social policy. But in the absence of mass participation, in the absence of a purely democratic egalitarian society, the social-science community provides, however furtive, however weak, a staff of rationality.

If one cannot believe in social science as a higher rationality, then all is lost. At such a point one must take recourse in irrationality, or violence, or some kind of behavior that excludes the possibility of corrective measures entirely, in or out of the bubble.

Thus, when we examine the problem of social-science numbers (and we are in a world of relatively small numbers), we are at least confronting the possibility of a common discourse, a common rationality, a common framework. To deny social science entirely, to go outside of it entirely, to put a plague on the house of society, or the house of the polity, does not come to terms with the structure of society. Within that larger structure, however weak, however peculiar, however at times distasteful, social scientists do represent some voice for discursive reason, some kind of pacification of barbarism and decency in human behavior.

An observation made by the sociologist Guy B. Johnson some forty years ago is well worth recalling. It was a time when it seemed as though the apocalypse was upon us, and nothing short of storming the gates of heaven or surrendering to the fires of hell made sense. "Personally," he wrote, "I should rather help to capture the foothills which have to be captured sooner or later than merely to point out the distant peak and urge my comrades to storm it at once! I, too, can see the peak, but I see no particular virtue in starting an association of peak-gazers."[6] Thus spoke a good social scientist.

Lest one imagine that the Balkanization phenomenon described herein is unique to the social sciences, we should be reminded of the situation in applied mathematics with respect to engineering. The "father" of applied mathematics, Theodore von Karman, describes contemporary conditions in the United States as follows: "American engineers are organized in separate societies. Mechanical, civil, electrical, aeronautical, and automotive engineers have their own organizations, and very little contact exists between them." A recent article amplifying this situation of the 1940s notes that "each engineering society had a separate journal. Applied mathematics issues, essentially interdisciplinary papers, had to find space in existing journals. From this intellectual atomization emerged new publications and new organizational forms."[7]

It must also be frankly noted that the reasons such Balkanization was overcome had less to do with the inner logic of mathematics of engineering than with the external compelling force of World War II. One must be careful not to exaggerate the policy proclivities or potentials of the social sciences. Past growth was largely determined by the needs of war and politics—no different from the situation in the physical sciences. Under what circumstances do some disciplines grow and others vanish? The social sciences are a function not simply of our inner scientific histories, but also of the larger histories of nations and peoples.

Robert Nisbet has caught the spirit of this fundamental dilemma when he noted that the classical antitheses, which for a hundred years have provided impetus for sociological generalizations, have simply run dry. "Community-society, authority-power, status-class, and sacred-secular all have vitality so long as the substantive equivalents have reality and relevance. It is like the distinction between state and society or between rural and urban: good so long as substantive referents are still existent, still imperative in their reality, but of diminishing significance, even illusory, once they are gone." Nisbet properly concludes by noting that the "tidal movements of change that up until now have given significance to those antitheses and their numberless corollaries have all too clearly reached a stage of completeness that cuts much of the empirical ground out from under the antitheses."[8] Thus, our understanding of the inner world of professional social life, to be given renewed meaning, must once again come to terms with real histories of real societies.

10

Human Life, Political Domination, and Social Science

If the first step toward a reconstruction of social research is to examine the actual conditions of professional life and academic ideologies, then the next step must surely be to locate a broad field of study—one that enlists the aid of empirical studies to normative issues. As I tried to demonstrate in *Taking Lives*, one critical frame of reference (although not necessarily the only such field) is the social meaning of life and death. I do not imply by this some course in the biology of the life-cycle, but rather the study of the mechanisms and purposes in the arbitrary termination of life in the name of state power, military necessity, racial purity, ethnic cleansing, or religious holy war.

Call it what one might, the examination of genocide, far more than the study of homicide, offers broad insights into a mass phenomenon that defines the twentieth century with as much force in negative terms as, say, the emergence of electronics does in positive terms. It also brings together quantitative and qualitative considerations of the utmost significance. We turn to this as a "case" study, not for the purpose of exhausting what has become a veritable cultural industry but to recognize in genocide that special and rare opportunity to reaffirm the fundamental moral purpose of social science without giving way to the belated, futile, and even insincere moral judgment of ideology.

The subject of genocide in general and the Holocaust in particular threatens to become a growth industry within the Western cultural apparatus. Books, plays, and television dramatizations on the subject pour forth relentlessly. Sometimes they are presented soberly, other times scandalously; but all are aimed at a mass market unfortunately more amazed than disturbed by their implications. There is danger in the massification of holocaust studies. Western culture is inclined to adopt fads; even holocaust studies may become a moment in commercial time—interest in them may decline as well as grow, and certainly peak out. Their residual debris will probably be summarized in musical comedy; we have already seen examples of this in *The Lieutenant* (Lieutenant Calley) and multiple versions of *Evita* (Eva Perón). Peter Weiss's play *The Investigation* led one commentator to suggest that the major character in the play, in order to elicit shock from the audience, read lines "as if he were saying: 'Let's hear it for genocide.'"[1] This may be a precusor of things to come.

One of the least attractive features of "post-holocaust" studies is the effort of a few to monopolize the field. As a consequence, a linguistic battle looms among survivors over which exterminations even deserve the appellation "holocaust" (the total physical annihilation of a nation or people). Such a bizarre struggle over language remains a grim reminder of how easy it is for victims to challenge each other and how difficult it is to forge common links against victimizers.[2] I do not wish to deny Jewish victims of the Nazi Holocaust the uniqueness of their experience. But there are strong elements of continuity as well as discontinuity in the process of genocide. It is the task of honest social research to take the measure of this emotive field of study—not simply to settle the question of genocide, but to appreciate how the issue of life and death is at the center of a social research agenda.

Writing with compelling insight, Elie Wiesel personifies the mystic vision of the Holocaust. Those who lived through it "lack objectivity," he claims, while those who write on the subject but did not live through it must "withdraw" from the analytic challenge "without daring to enter into the heart of the matter."[3] More recently, it has been suggested that "for Jews, the Holocaust is a tragedy that cannot be shared," and "it may be unrealistic or unreasonable or inappropriate to ask Jews to share the term holocaust. But it is even more unreasonable and inappropriate not to find a new name for what has taken place in Cambodia."[4] Since what took place in both situations is a holocaust—from the demographic point of view—we need not invent new terms to explain similar barbaric processes. Those who share a holocaust share a common experience of being victim to the state's ruthless and complete pursuit of human life-taking without regard to individual guilt or innocence. It is punishment for identification with a particular group, not for personal demeanor or performance. This is not a theological but an empirical criterion. To seek exclusivity in death has bizarre implications. The special Jewish triumph is life. All too many peoples—Jews, Cambodians, Armenians, Paraguayans, Indians, Ugandans—have shared the fate of victims to engage in divisive squabbles about whose holocaust is real or whose genocide is worse.

Those who take an exclusive position are engaging in moral bookkeeping, in which only those who suffer very large numbers of deaths qualify. Some argue that the six million deaths among European Jews is far greater than the estimated one million deaths among Armenians. However, the number of Armenian deaths as a percentage of their total population (50 percent) is not much lower than the percentage of Jewish losses (60 percent). Others contend that to few Ugandans or Biafrans were killed to compare that situation to the Holocaust, yet here, too, tribal deaths in percentage terms rival the European numbers. In certain instances, high death rates (for example, approximately 40 percent of all Cambodians or three million out of seven million) are indisputable; then one hears that such deaths were only random and a function of total societal disintegration. Yet it has been firmly established that such violence was targeted against intellectuals, educators, foreign-born, and literate people—in short, the pattern was hardly random; anyone who could potentially disrupt a system of agrarian slave labor flying under communist banners was singled out and eliminated. Even making the definition of "holocaust" a matter of percentages risks creating a morality based solely on bookkeeping.

There is a need to reaffirm the seriousness of the subject; and in so doing, the worthwhileness of social-science analysis. The problem of genocide must be rescued from mass culture. It must not be returned to academic preserves, but it must be made part and parcel of a general theory of social systems and social structures. The positions that I would like to discuss, examine, and criticize have perhaps best been articulated by the theologian Emile C. Fackenheim[5] and the anthropologist Leo Kuper.[6] In some curious way, they represent the extremes that must be overcome if an integrated approach to the study of genocide is to become a serious subject for scientific analysis. On the one hand, Fackenheim speaks with a thunderous theological certitude that approaches messianic, or at least prophetic, assuredness. At the other end is Kuper, who is extremely modest in his approach, to the point where some fundamental distinctions between severe strife and mass destruction are entirely obliterated. This is not to suggest that the truth lies somewhere in the middle but rather that the need for a social-scientific standpoint in the study of genocide may convince all to move to a higher ground in this area—an area of research that has truly replaced economics as the dismal science.

Fackenheim's propositions have come to represent the main trends in the theological school of Holocaust studies. They carry tremendous weight among mass-culture figures for whom theological sanction provides legitimation to their endeavors and respite from critics.[7] Professor Fackenheim does not remotely intend his views to become part of mass culture; quite the contrary. His eight propositions distinguishing the Holocaust in particular from genocide in general represent a tremendous effort to transcend journalistic platitudes, to move beyond an articulation of the banality of evil to the evil of banality. This deep respect for Fackenheim registered, it must also be said that an alternaive perspective, a social-science framework, is warranted.

Fackenheim presents his eight propositions with direction and force. A general theory of genocide and state power, which accounts for the specifics of the Holocaust, can have no better base line.

One: *The Holocaust was not a war. Like all wars, the Roman War against the Jews was over conflicting interests—territorial, imperial, religious, other—waged between parties endowed, however unequally, with power. The victims of the Holocaust had no power. And they were a threat to the Third Reich only in the Nazi mind.*

The Holocaust *was* a war; but one of a modern rather than medieval variety. Earlier wars redistributed power by military means. Genocide redistributes power by technological as well as military means. Robert Lifton stated the issue succinctly.

The word holocaust, from Greek origins, means total consumption by fire. That definition applies, with liberal grotesqueness, to Auschwitz and Buchenwald, and also to Nagasaki and Hiroshima. In Old Testament usage there is the added meaning of the sacrifice, of a burnt offering. That meaning tends to be specifically retained for the deliberate, selective Nazi genocide of six million Jews—retained with both bitterness and irony (sacrifice to whom for what?). I will thus speak of the Holocaust and of holocausts—the first to convey the uniqueness of the

Nazi project of genocide, the second to suggest certain general principles around the totality of destruction as it affects survivors. From that perspective, the holocaust means total disaster: the physical, social, and spiritual obliteration of a human community.[8]

The precedent for this war against the Jews was the Turkish decimation of the Armenian population. Like the Nazis, the Ottoman Empire did not simply need to win a war and redistribute power; they had an overwhelming amount of power to begin with.[9] A war of annihilation is a war. To deny the warlike character of genocide is to deny its essence: the destruction of human beings for predetermined nationalist or statist goals.

The holocaust is also modern in that it is an internal war, waged with subterfuge and deception by a majority with power against an internal minority with little power. Here, too, the Armenian and Jewish cases are roughly comparable. Although one can talk of genocide in relation to the bombing of Hiroshima and Nagasaki, genocidal conflict involves internal rather than external populations. But this is an unambiguous point on the nature of war rather than a denial of the warlike nature of the holocaust *per se.*

The victims of the Holocaust did have a certain power: they represented a threat to the Nazi Reich. The Jew as bourgeois and the Jew as proletarian represented the forces of legitimacy and revolution in Weimar Germany. They had modest positions in universities, in labor, and in industry. Regarding state power itself, where hardly any Jews held positions of influence, they were powerless. Jews were locked out from the German bureaucratic apparatus much as the Turkish Beys locked out Armenians from the administrative apparatus, except to use them in a Quisling-like manner. The Jews posed a threatening challenge to the legitimacy of the Nazi regime.

Two: *The Holocaust was not part of a war, a war crime. War crimes belong intrinsically to wars, whether they are calculated to further war goals, or are the result of passions that wars unleash. The Holocaust hindered rather than furthered German war aims in World War II. And it was directed, not by passions, but rather by a plan conceived and executed with methodical care, a plan devoid of passion, indeed, unable to afford this luxury.*

This argument rests on a peculiar and misanthropic rendition of the Hilberg thesis. The Holocaust internments did hinder the Nazi war effort in the limited sense that troop transportation took second place to transporting Jews. But in the longer, larger perspective, there were advantages. Slave labor was itself an advantage; unpaid labor time was useful. The expropriation of goods and materials was an economic gain for the Nazi Reich. People were liquidated at marginal cost to the system. The gold taken from extracted teeth became a proprietary transfer.[10] Fackenheim questions whether war goals were furthered by the Holocaust; this is not simply answered. As a mobilizing device linking military and civil sectors of the populaion, the Holocaust enhanced war ends. The Nazi attempt to exterminate the Jews was motivated by passion, as evidenced by the fact that even troop movements to the Russian front took second place to transporting Jews to internment camps.

Hilberg makes clear the direct collusion of the German Wehrmacht and the German Reichsbahn with respect to the systematic deportation of Jews and the front-line servicing of the armed forces. The management of the German railroad illustrates how irrationality can become rationalized, how a "true system in the modern sense of the term" was employed for the unrelenting destruction of human lives. As Hilberg notes, to the extent that the technicalization of mass society was exemplified by the transportation network, such human-engineering considerations cannot be viewed as ancillary.

> It illuminates and defines the very concept of "totalitarianism." The Jews could not be destroyed by one Fuhrer on one order. The unprecedented event was a product of multiple initiatives, as well as lengthy negotiations and repeated adjustments among separate power structures, which differed from one another in their traditions and customs but which were united in their unfathomable will to push the Nazi regime to the limits of its destructive potential.[11]

The question of passion is a moot point at best; undoubtedly there was a collective passion undergirding the conduct of the Holocaust. It was not simply a methodical event.

Fackenheim, and many other theologians, overlooked parallels in the pursuit of a genocidal state following defeat. After the Turkish defeat at the hands of Bulgaria in 1912, the most massive genocide against Armenians occurred. After the Nazi defeat at Stalingrad in 1943, the most massive destruction of Jews ensued. Whatever the vocabulary of motives—fear of discovery, of reprisal, or of judgment—the use of state-sanctioned murder to snatch national victory from the jaws of international defeat is evident. The Nazis may have lost the war against the West, but they achieved their primary objective: the destruction of the Jewish people of Germany.

The largest part of European Jewry was destroyed *after* Germany had, in effect, lost the war. When the major object of the war, defeating the Allied powers, was no longer feasible, the more proximate aim, destroying the Jewish people, became the paramount goal. War aims have manifest and latent elements. The manifest aim was victory in the war, but the latent aim was defeat of the internal "enemy," the Jews. The decimation and near-total destruction of the Jewish population might be considered the victory of the Third Reich in the face of the greater defeat they faced by the end of Stalingrad.

Three: *The Holocaust was not a case of racism, although, of course, the Nazis were racists. But they were racists because they were anti-Semites, not anti-Semites because they were racists. (The case of the Japanese as honorary Aryans would suffice to bear this out.) Racism asserts that some human groups are inferior to others, destined to slavery. The Holocaust enacted the principle that the Jews are not of the human race at all but "vermin" to be "exterminated."*

Here Fackenheim represents a considerable body of thought. But the Holocaust *was* a case of racism. It is not a question of which comes first, anti-Semitism or racism; that taxonomical dilemma is secondary. Assignment of special conditions of life and work to Jews implies what racism is all about: the assumption of inferiority and superiority leading to different forms of egalitarian outcomes. Ultimately, racism is not about institutionalizing inferiority or superiority but about denial of the humanity of those involved. Jewish vis-à-vis Aryan physical characteristics were studied by German anthropologists to prove that there was such a thing as race involved. These stereotypes were the essence of European racism, as George Mosse has fully documented in his recent work.

> Racism had taken the ideas about man and his world which we have attempted to analyze and directed them toward the final solution. Such concepts as middle-class virtue, heroic morality, honesty, truthfulness, and love of nation had become involved as over against the Jew; the organs of the efficient state helped to bring about the final solution; and science itself continued its corruption through racism. Above all, anthropology, which had been so deeply involved in the rise of racism, now used racism for its own end through the final solution. Anthropological studies were undertaken on the helpless inmates of the camps. Just as previously non-racist scientists became converted by the temptation to aid Nazi eugenic policies so others could not resist the temptation to use their power over life and death in order to further their anthropological or ethnographic ambitions.[12]

The fact that American racism has a clear-cut criterion based on skin color does not mean that the physical and emotional characteristics attributed to Jews were less a matter of racism than the characteristics attributed to American blacks. To deny the racial character of the Holocaust is to reject the special bond that oppressed peoples share. To emphasize distinctions between peoples by arguing for the uniqueness of anti-Semitism carries with it profound risks: it reduces any possibility of a unified political and human posture on the meaning of genocide or of the Holocaust. The triumphalism in death implicit in this kind of sectarianism comes close to defeating its own purpose.

> Four: *The Holocaust was not a case of genocide although it was in response to this crime that the world invented the term. Genocide is a modern phenomenon for the most part. In ancient times human beings were considered valuable, and were carried off into slavery. The genocides of modern history spring from motives, human, if evil, such as greed, hatred, or simply blind xenophobic passion. This is true even when they masquerade under high-flown ideologies. The Nazi genocide of the Jewish people did not masquerade under an ideology. The ideology was genuinely believed. This was an "idealistic" genocide to which war aims were, therefore, sacrificed. The ideal was to rid the world of Jews as one rids oneself of lice. It was also, however, to "punish" the Jews for their "crimes," and the crime in question was existence itself. Hitherto, such a charge had been directed only at devils. Jews had now become devils as well as vermin. And there is but one thing that devils and vermin have in common: neither is human.*

Here Fackenheim has a problem of logical contradiction. First we are told that the Holocaust is not a case of genocide; and then we are reminded of the Nazi genocide of the Jewish people. But more significant is the contradiction within this framework, an inability to accept the common fate of the victims. Whether they are Japanese, Ugandans, Gypsies, Cambodians, Armenians, or Jews, their common humanity makes possible a common intellectual understanding. Insistence upon separatism (that the crime was Jewish existence—and this makes the Jewish situation different from any other slaughter, whatever its roots) contains a dangerous element of mystification. It represents a variation of the belief in "chosenness," converting it from chosen to live God's commandments into chosen for destruction. This approach is dangerously misanthropic. It misses the point that being chosen for life may be a unique Jewish mission but being slected for death is common to many peoples and societies.

The description of Jews as devils was not the essence of Nazi anti-Semitism; it was only the rhetoric of a regime bent on the uses of state power to destroy a powerful people. The Ayatollah Khomeini and other Iranian clerics constantly refer to Americans as devils in the same way. The essence of the Jewish problem for Nazism was the Jew as a political actor and beyond that, the Jew as a cosmopolitan, universalistic figure in contrast to fascist concepts based on nationalism, statism, and particularism. The Jewish tradition of social marginality, of reticence to participate in nationalistic celebrations, makes anti-Semitism a universal phenomenon, as characteristic of the French as of the Russians. The special character of Jewish living cannot be easily converted into the special nature of Jewish dying. Dying is a universal property of many peoples, cultures, and nations.

> Five: *The Holocaust was not an episode within the Third Reich, a footnote for historians. In all other societies, however brutal, people are* punished for doing. *In the Third Reich, "non-Aryans" were "punished" for being. In all other societies—in pretended or actual principle, if assuredly not always in practice—people are presumed innocent until proved guilty; the Nazi principle presumed everyone guilty until he had proved his "Aryan" innocence. Hence, anyone proving or even prepared to prove such innocence was implicated, however slightly or unwittingly, in the process which led to Auschwitz. The Holocaust is not an accidental by-product of the Reich but rather its inmost essence.*

Any response to this proposition must acknowledge the basic truths of the first part of the statement. The Holocaust was not merely a passing moment within the Third Reich. It did not occur in other fascist countries, like Italy, for example, where death itself was alien to the Italian culture, where the survival not only of Jews but also of communists was tolerated and even encouraged. Antonio Gramsci's major works were written in a prison that had been converted into a library by his jailers. The nature of national culture is a specific entity. The Italian people, the Turkish people, and the German people all had a distinctive national character. Social analysts do not discuss this kind of theme in public. It is not fashionable; we have become even a bit frightened of the concept of national character. Any notion of national character carries with it the danger of stereotypical thought. But how else can we understand these phenomena? How can we understand the character of reaction, rebellion, and revolution in Turkey without understanding Turkish character, especially the continuity of that kind of character, across political regimes, in the moral bookkeeping of development.

Ascribing guilt through proving innocence fits the framework of the Nazi ideology. But to construct a general theory of historical guilt may have pernicious consequences, in which the sins of the fathers are bequeathed to the children and further offspring. That the Holocaust was "inmost essence" makes it difficult to get beyond phylogenic memories, beyond a situation in which a society might be viewed as having overcome its racism. When guilt is generalized, when it no longer is historically specific to social systems and political regimes, then a kind of irreducible psychologism takes intellectual command, and it becomes impossible to stipulate conditions for moving beyond a genocidal state. The Holocaust becomes part of a rooted psychic unconsciousness hovering above the permanently contaminated society. To be sure, the Holocaust is the essence of the Third Reich. However, such an observation is not necessarily the core question. Does the destruction of the Jews follow automatically upon a nation that is swallowed up by the totalitarian temptation? Is anti-Semitism the essence of Russia as is now claimed? Does the existence of anti-Semitism prove a theory of totalitarian essence?

The uncomfortable fact is that genocide is the consequence of certain forms of unbridled state power. But whether anti-Semitism or other forms of racial purity are employed depends on the specific history of oppressor groups no less than oppressed peoples. States that demonstrate their power by exercising their capacity to take lives may be termed "totalitarian." Totalitarianism is the essence of the genocidal process. This in itself provides an ample definition. If the Holocaust is unique to the Third Reich, the question of genocide loses any potential for being a general issue common to oppressive regimes. It is parochial to think that the Third Reich somehow uniquely embodied the character of the Holocaust, when we have seen since then many other societies adopt similar positions and policies toward other minorities and peoples.

> Six: *The Holocaust is not part of German history alone. It includes such figures as the Grand Mufti of Jerusalem, Hajj Amin al-Husseini, who successfully urged the Nazi leaders to kill more Jews. It also includes all countries whose niggardly immigration policies prior to World War II cannot be explained in normal terms alone, such as the pressures of the Great Depres-*

sion or a xenophobic tradition. Hitler did not wish to export national socialism but only anti-Semitism. He was widely successful. He succeeded when the world thought that "the Jews" must have done something to arouse the treatment given them by a German government. He also succeeded when the world categorized Jews needing a refuge as "useless people." (In this category would have been Sigmund Freud had he still been in Germany rather than in England; Martin Buber had he not already made his way to the Yishuv [Palestine].) This was prior to the war. When the war had trapped the Jews of Nazi Europe, the railways to Auschwitz were not bombed. The Holocaust is not a parochial event. It is world-historical.

Curiously there is no mention of any other kind of history. Is, for example, the genocide of the Armenian people part of world history, or is it simply part of Turkish history? This is a complicated point; at the risk of sounding impervious to moral claims, one has to be history-specific if anything serious is to emerge. If one blames the whole world for what took place at Vin, one can construct such a theory. But it is more pertinent, more appropriate, more pointed to blame the Turks and not the universe and to blame the Germans and not the whole world for the Holocaust. The issue is implementation, not rhetoric. The issue is neither the Grand Mufti nor the insecurities of Ambassador Morgenthau. Rather, social research needs to establish how state power moves from the regulation to extermination of people.

As Gideon Hausner reminds us,[13] as late as April 1945, when the Soviets were penetrating Berlin for the final assault and when Hitler was imprisoned in his bunker, his last will and testament concluded by enjoining "the government and the people to uphold the racial laws to the limit and to resist mercilessly the poisoner of all nations, international Jewry." Hausner makes it plain that national socialism was an international movement, the lynchpin of which was anti-Semitism. Fackenheim presumes World War II was all about anti-Semitism, but at a more prosaic level, it was about conquest. There was a Nazi government in the Ukraine; there was a Nazi government in Norway; there was a Nazi government in Romania; there was a Nazi government in Yugoslavia—all these regimes were exported. The idea that Hitler was not interested in exporting national socialism is curious. It would be more appropriate to note that to wherever national socialism was exported, so, too, did anti-Semitism follow. However, in conditions where the Jewish population was not a factor, Nazism still sought to establish a political foothold—either with or without direct military aggression. The relation between national socialism as an ideology and anti-Semitism as a passion is one that the Nazis themselves were hard put to resolve. The linkage between the ideology and the passion, which seems so close in retrospect, was by far less an articulated policy than a felt need in the earlier states of the Nazi regime.

Fackenheim slips in a subtle point that the Jews were "trapped" in Europe. But the Jews were not trapped in Europe. They were of Europe and had been of Europe for a thousand years. One of their dilemmas is one rendered in almost every history: those who are to be exploited or annihilated overidentify with their ruling masters. The Jews of Europe were entirely Europeanized. Only a small fragment remained outside the framework of Europeanization. The great divide of German and Russian Jews was participation in European nationalism, identification with enlightenment. Fackenheim's idea that the Jews were trapped in Europe is a clever misreading of the facts. The added horror of the Holocaust is that it happened to a people who were endemic to that part of the world.

Seven: *The Jews were no mere scapegoat in the Holocaust. It is true that they were used as such in the early stages of the movement. Thus Hitler was able to unite the "left" and "right" wings of his party by distinguishing, on the left, between "Marxist" (i.e., Jewish) and "national" (i.e., "Aryan") "socialism" and, on the right, between* Raffendes Kapital *(rapacious, i.e., Jewish capital) and* Schaffendes Kapital *(creative, i.e., "Aryan" capital). It is also true that, had the supply of Jewish victims given out, Hitler would have been forced (as he once remarked to Herman Rauschning) to "invent" new "Jews." But it is not true that "The Jew [was] . . . only a pretext for something else. So long as there were actual Jews, it was these actual Jews who were the systematic object of ferreting-out, torture, and murder. Once, at Sinai, Jews had been singled out for life and a task. Now, at Auschwitz, they were singled out for torment and death.*

The difficulty with this exclusivist formula is that while Jews were singled out, so, too, were Gypsies, Poles, and Slavs. Hitler's appeal was to state power, not to unite Right and Left, not to unite bourgeoisie and proletariat, but to make sure that the bourgeoisie and the proletariat of Germany were purified of Jewish elements. If one considers the national aspects of the Third Reich rather than the mystical aspects of Jewish destruction, this becomes a lot easier to fathom. German-Jews were concentrated in the bourgeoisie and proletariat—in leftist socialist politics and in high bourgeois economics. Liquidation of the Jews enabled the German bureaucratic state to manage the bourgeoisie and proletariat of Germany without opposition.[14] The destruction of socialism was attendant to the destruction of the Jews. Without socialist opposition, the German proletariat was an easy prey to Third Reich massification. The first two legislative acts of the Third Reich were bills on labor, on work and on management. The liquidation of the Jewish population, within both the bourgeoisie and the proletariat, permitted the Nazis to consolidate state power. The Holocaust, from a Nazi standpoint, was an entirely rational process, scarcely a singular act of mystical divination. It was the essential feature of Nazi "domestic" policy in the final stages of the Third Reich.

Eight: *The Holocaust is not over and done with. Late in the war Goebbels (who, needless to say, knew all) said publicly and with every sign of conviction that, among the peoples of Europe, the Jews alone had neither sacrificed nor suffered in the war but only profited from it. As this was written, an American professor has written a book asserting that the Holocaust never happened, while other Nazis are preparing to march on Skokie, in an assault on Jewish survivors. Like the old Nazis, the new Nazis say two things at once. The Holocaust never happened; and it is necessary to finish the job.*

On this point, Fackenheim is on sound ground. The Holocaust did happen and could happen again. But as recent events in parts of Europe as well as Africa make clear, it is now more likely to happen to peoples other than Jews or Armenians. It was more likely to happen to Ugandans, and it did; to Cambodians, and it did; to Paraguayans, and it did; to Biafrans, and it did. It is correct to say that the Holocaust is not over and done with. But it is not over and done with because there are other peoples victimized by the very model created by the Turkish and Nazi genocides.

It is important not to fit "peoplehood" into social theories; theories must fit the realities of people. If the restoration of human dignity is to become a theme for social research, it becomes imperative to understand the unified character of genocide—the common characteristics of its victims. Alliances of victims and potential victims are needed to resist all kinds of genocide. To narrow the term on the basis of universalism, triumphalism, or separatist orientations is self-defeating. If there is to be any political consequence of research into genocide, if the victim groups are to do more than pay for annual memorials and remembrances, then to engender understanding of the unity needed to confront state oppression must be a paramount goal; otherwise, little will have been accomplished and nothing will have changed.

Although my analysis has sharply demarcated theological from sociological viewpoints, it should be appreciated that Jewish religious thought is itself far from unanimous on the special nature of the Holocaust. Orthodox segments in particular have cautioned against an overly dramaturgical viewpoint, urging instead a position in which the Nazi Holocaust is but the latest monumental assault on the Jewish people; one that is to be neither ignored nor celebrated but simply understood as part of the martyrdom of a people. In a recent essay, Helmreich has finely caught the spirit of this "strictly orthodox" view—which may be shared by larger numbers than either the mystifiers or the celebrationists may recognize.

He notes that his orthodox wing rejects paying special homage by singling out the victims of the Holocaust on both philosophical and practical grounds.

> In their view, the Holocaust is not, in any fundamental way, a unique event in Jewish history, but simply the latest in a long chain of anti-Jewish persecutions that began with the destruction of the Temple and which also included the Crusades, the Spanish Inquisition, attacks on Jews led by Chmielnicki, and the hundreds of pogroms to which the Jewish community has been subjected to over the centuries. They do admit that the Holocaust was unique in scale and proportion but this is not considered a distinction justifying its elevation into a separate category.[15]

Helmreich goes on to note that the ethical problem, in the view of orthodox believers, is the same if one Jew is murdered or if six million meet such a fate. Since Judaism is a *Gemeinschaft*, a community of fate, the sheer volume killed, while awesome, does not in itself transform a quantitative event into a unique qualitative phenomenon.

The significance of this minority theological report is to call attention to the fact that in the problem of the Holocaust, while there are some strong clerical/secular bifurcations, there are also patterns cutting across disciplinary boundaries. For example, certain sociological lessons can be drawn from the Holocaust: the breakdown in egalitarian revolutions of the nineteenth century, the subtle abandonment of the Palestinian mandate after the Balfour Declaration, and the lofty assertion followed by a total revocation of Jewish minority rights in the Soviet Union. For orthodoxy, the Holocaust is more a function of the breakdown of Jewish solidarity than of any special evils of the German nation or the Nazi regime.

The sociological view attempts to transcend sectarian or parochial concerns and develop a cross-cultural paradigm that would permit placing the *Holocaust* into a larger perspective of *genocide* in the twentieth century—rather than see the former as entirely distinctive and the latter as some weaker form of mass murder. With the liquidation of roughly 40 percent of the Cambodian population, even the quantitative indicators of the Nazi Holocaust have been approached in at least one other situation. In the past, it has been argued that genocide of other peoples—Armenians, Ugandans, Paraguayans, Indians—has been too random and sporadic to be termed a holocaust. It has also been claimed that atomic attacks on Hiroshima and Nagasaki were highly selective and refined military targets and not efforts at the total destruction of a people. Whatever the outcome of such contentions, the Cambodian case would indicate the risks in vesting too much intellectual capital in the sheer numbers involved—although that is clearly a factor to be contended with.

Having argued thusly, let me note that qualitative differences do exist between the Jewish Holocaust and *any* other form of genocide. First, there is the systematic rather than the random or sporadic nature of the Holocaust: the technological and organizational refinement of the tools of mass slaughter which ultimately reduced all morality to a problem of human engineering as to the most effective method for destroying and disposing of large numbers of people by the fewest cadres possible in the shortest amount of time. Second, there was an ideological fervor unmatched in any other genocide. So intent were the Nazis on their policy of extermination of Jews that they dared contact other nations, especially Axis powers and neutral countries, to repatriate Jews back to Germany to suffer the ultimate degradation. Third, genocide against the Jewish people represented and rested upon a national model of state power: the purification of the apparatus of repression by a total concentration of the means of destruction in a narrow military police stratum unencumbered by considerations of class, ethnicity, gender, or any other social factors affecting Nazi response to non-Jewish groups. The liquidation of plural sources of power and authority made easier, indeed presupposed, the total liquidation of the Jewish population.

With all these inner disputations and disagreements accounted for, there are still those who—too guilt ridden to face the monstrous consequences of the Holocaust against Jews in particular and victims of genocide as a whole—have chosen the path of evading reality. An isolated voice like that of Arthur R. Butz[16] is now joined in a quasi-intellectual movement with all the paraphernalia of historical scholarship[17]—the beginning of a massive denial of a massive crime. Denials of gas chambers, rejection of photographic evidence, and equation of indemnification of the victims with Zionist beneficiaries are all linked to the rejection that the Holocaust ever occurred. The Nazi "revisionists" dare not speak of Nazism but only of national socialism, not of Germany under Hitlerism but only of a Third Reich. The Nazi era is even spoken of in remorseful terms: "Overwhelming British, American, and Soviet forces finally succeeded in crushing the military resistance of a Germany which they accorded not even the minimum of mercy."[18] Pity the poor victim!

Even the neo-Nazi "intelligentsia" does not deny mass murder but only the numbers murdered.[19] It is not supposedly six million (then what number is it?). No matter, those massacred were Zionists, Communists, or a hyphenated variety of the two (Jewish-Bolsheviks)—any euphemism for Jews that denies a special assassination of Jews as a people. The need for exacting scholarship—the sort that has begun to emerge—with respect to all peoples victimized merely on the basis of their existence is not a matter of litanies and recitations but of the very retention of the historical memory itself. The scientific study of genocide is not a matter of morbid fascination or mystic divination but of the need to assert the historical reality of collective crime. Only by such a confrontation can the social sciences help assign moral responsibility for state crimes, even if they cannot in or of themselves prevent future genocides from taking place.

With due weight given to the different traditions involved in the theological and sociological arguments concerning genocide, they do have a strong shared-value commitment to the normative framework in which greater emphasis is placed on the protection of life than on economic systems or political regimes.[20] Both traditions are committed, insofar as their dogmas and doctrines permit, to the supreme place of life in the hierarchy of values. This is no small matter. Nazism witnessed the breakdown of religious and scientific institutions alike; and those that could not be broken down were oftentimes simply corrupted, as in decadent and exotic notions of a Teutonic Church and the equally ludicrous belief in an Aryan Science. In the larger context of world history, in the wider picture of centuries-old barbarisms, we bear witness not to a warfare of science versus theology but rather a shared collapse of any sort of normative structure in which either could function to enhance the quality or sanctity of life.

Leo Kuper is professor emeritus of anthropology at the University of California, Los Angeles, born and banned in South Africa, and the author of several excellent monographs on social stratification and race relations in African contexts. He is a good man writing on an awful subject who has produced, unfortunately, a mediocre book. This may seem to be a severe judgment about an effort undertaken with deep commitment and intellectual integrity; but his American publisher, who insisted on counting *Genocide* as "the first systematic treatment of the subject," which it is clearly not, must be held partially accountable. The author himself (like the original British publisher) is far more modest, describing his work as a series of case studies in domestic genocides; that is, those internal to a society. But the fault is, I am afraid, not only that of overly enthusiastic advertising copy. The author manages to skirt just about every major issue which has arisen in the field of investigation: the relationship between the Holocaust and genocides in general; the relationship between civil conflict and state destruction; and the reasons for the ineffectiveness of international peace-keeping agencies in reducing genocide. The last omission is particularly glaring since Professor Kuper sets this as an essential task.

On a different level, however, it is an excellent basic text, especially for individuals who are not familiar with the subject of genocide. Definitions are invariably fair-minded and essentially sound; the appendices, especially on U.N. resolutions and areas of backsliding (such as its attitude toward the Turkish genocide against Armenians), are particularly revealing. The role of the United Nations (or better, its lack thereof) has often been talked about but little understood. At such descriptive levels, the book provides a welcome contribution. The cases selected are, for the most part, helpful and show a keen sense of the magnitude of the problems of genocide. When we talk in terms of roughly eight hundred thousand Armenians, six million Jews, and three million Cambodians, we have clear-cut examples of an enormous portion of a national population decimated by the authorities, giving a sober reminder that our century hovers dangerously between creativity and destruction.

Kuper's anthropological problems are less those of sentiment than of method. Equating such phenomena as civil strife between Catholics and Protestants in Northern Ireland with the destruction of German Jewry or the destruction of urban Cambodia just does not work. Even the author acknowledges that in Northern Ireland, victims have been numbered in the hundreds over a long stretch of time, whereas in most clear-cut cases of genocide the numbers destroyed are in the millions. Then there is the too simplistic equation of civil war and genocide. Equating the Nigerian Civil War, or even the struggle against Apartheid in South Africa, with cases of undisputed genocide blurs rather than clarifies what genocide is about; namely, the vast, near-total destruction of large numbers of noncombatants innocent of any specific crime. Further, faulting the United Nations as the source of the failure to control genocide is unconvincing since, as the author is at pains to explain, this organization is primarily a composite of nations and not in itself a sovereign power. Underneath the demand to strengthen the United Nations is an implicit assumption that nationalism should be weakened, something that clearly neither has nor is likely to take place—certainly not under the aegis of the United Nations.

The problem of genocide is not a new one, and the need for a literature to move beyond horror and into analysis becomes increasingly critical. It is risky to equate genocide with arbitrary death. Two examples which Leo Kuper has given illustrate a problem rather than indicate a solution. He raises, for example, the case of India during the partition, in which Hindus and Moslems constituted majorities in different parts of the country, each group with the capacity to engage freely in what he calls reciprocal genocidal massacre. However terrible and tragic that mutual destruction was, to speak of it as genocidal, in a context of religious competition and conflict, risks diluting the notion of genocide and equating it with any conflict between national, religious, or racial groups. This error also appears in his analysis of Northern Ireland, where Protestants and Catholics engage in the meanest and most dangerous kinds of assaults on one another. If deaths since 1920 were tallied, they would hardly be above ten thousand, surely a terrible human loss and with equal surety an indicator of the risks involved in Kuper's notion that the removal of the British presence and the withdrawal of the British army before a political solution in Ulster is achieved would invite bloodshed. One might indicate, as many leaders from both the Catholic and Protestant camps have, that the British presence is itself a source of violence and that the removal of the occupying power would overcome a major obstacle to resolution of the civil conflict. Whether this belief is correct or not, we are dealing here with the realm of political tactics and international relations, surely not with the area of genocide—unless we reduce the term to a fatuous notion of the cultural elimination of certain groups and ideologies.

Kuper also confounds legal identification between Apartheid and genocide in South Africa with the empirical problem: the place and condition of the blacks within South Africa. As the author himself well appreciates, there is a demographic restraint to annihilation. The black African population in South Africa grew from roughly 8,000,000 in 1945 to 19,000,000 in 1980. The Asian population grew from 285,000 to 765,000; and the white population, from 2,400,000 to roughly 4,400,000. The demographics alone indicate that genocide simply has not occurred. What may have happened is the fragmentation of the African population and the consequent denial of citizenship rights for most blacks. South Africa is also a classic case of exploitation of the majority by a racial minority, in a very specialized context. But it does not service the victims nor anyone else to present South Africa as a case of genocide—which implies the absolute destruction of a people; if not completely, then in such large numbers as to affect future survival potential as well as the present population.

Relativizing the issue of genocide particularly damages efforts to understand the Nazi Holocaust against the Jews. The major problem in such relativizing is that it completely fails to distinguish between the systematic, total, scientific engineering of death and the more random occurrences that are characteristic of other events. If others were to operate under the veil of anonymity as the Nazis did, they might also attempt a kind of "final solution." But whether that is so or not, the notion of the final solution is absent in the work of Professor Kuper. While I myself have argued against celebrating the exclusivity of death, one must take seriously differences between the total decimation of a population, reducing it to a remnant, and the selective, random elimination of political or religious opposition. The very concept of Holocaust fails to appear in Kuper's work and is mentioned only in relation to a book title. It is as if the author were consciously and deliberately attempting to relativize the Jewish case as one of many, consequently disregarding the specificity and peculiar characteristics involved in the Nazi Holocaust. This not only undermines the moral basis of Kuper's work but also weakens his appreciation of the full meaning of the Turkish assault on the Armenians. The latter was not merely an event that took place in the Ottoman Empire but is characteristic of the Kermalist democracy which followed. The genocide against Armenians, like the Holocaust against Jews, was special in its totality, in its movement beyond the boundaries of nationalism and rationalism. Both "cases" are not characteristic of any others—until we get to Kampuchean communism.

Underlying a failure to distinguish between genocide and civil strife, on the one hand, and genocide and such total destruction as the Holocaust, on the other, is a peculiar inability to distinguish between theory and action and, more specifically, an unwillingness to deal with culture, particularly German and Turkish cultures. Kuper, along with others, has spent a great deal of futile time on problems of ideology. There is sufficient confusion within Marxism and fascism to make one wary of this line of approach. Perhaps Marxism, in its acceptance of a theory of class polarization, yields to a Manichean vision of a world torn apart; but even a Marxism predicated on guilt by social origin may or may not translate into genocidal behavior. It certainly does in terms of the Gulag Archipelago and the years of Stalinism; and it may once again in such places as the former Yugoslavia, where doctrines of "ethnic cleansing" have been introduced.

Even within the fascist states of World War II, differences existed. The Nazis analogized European Jewry to a cancer that had to be excised and identified Jews with world conspiracy. However, fascism in Italy did not have the same genocidal potential. Even within provinces held by Japan during World War II, when Jewish enclaves came under Japanese dominion, the genocidal pattern did not obtain. Any comprehensive analysis of genocide must deal seriously with cultural canons which permit or forbid genocidal behavior. This total absence of analysis in the culture of peoples not given to genocide, or in that of those who were, seriously weakens Kuper's book. Ideology rather than culture is held responsible for genocidal behavior. This is a difficult thesis to prove. The developmentalism of the Kemalist regime in Turkey was absolutely at odds with Imperial notions derived from the Ottoman Empire, yet both democratic and antidemocratic forces within Turkey carried on genocide against the Armenians. The peasant egalitarianism of the Khmer Rouge did not spare the world a major genocide. Wherever one seeks an answer based on ideology, the same kind of confusion presents itself—an issue which Kuper unfortunately does not address.

It might well be the case that the United Nations did much less than it should have to curb genocide. Kuper argues that its capacity to curtail genocide, much less to prevent or punish atrocities, has been blunted. He gives several reasons for this laxity. First, the punitive procedures of the United Nations are weak; second, the United Nations is committed to the sanctity of state sovereignty; and third, the United Nations has established commissions to deal with complaints about human-rights violations which are themselves highly politicized and in the control of a clique of powerful nations whose vested interests are in stilling the voices of opposition. One could hardly argue with Kuper's analysis of the weaknesses of the United Nations system; but from an analytical point of view, it is an extremely thin reed upon which to hang an analysis of the problem of genocide. The sources of genocide are certainly not in the United Nations. Therefore, prospects for a solution are similarly not likely to be found in the United Nations. The limits of the organization are well understood. The social sciences must now focus on different kinds of national cultures and how attitudes toward law, justice, and punishment emerge in various countries and communities.

There is now a burgeoning literature on just these subjects. It might be possible to develop an early warning signal, a concern about problems of law and democratic order, that might limit the possibility of future genocides taking place. But if the genesis of the problem is not in a world organization, then it is hard to believe that the solution will be found there. The United Nations is itself the source of such intense national posturing and self-righteousness that its very existence ends in paradox: the strengthening of nationalism and the national idea.

From a public-policy viewpoint, the main task is to ensure the process of democratic development without inviting genocidal forms of exclusive racial or ethnic domination. This is easier said than done, for one must confront conflicting impulses of nationalist sentiment and minority rights. Even those societies with a nominal commitment to democratic order have responded with bigotry to the needs of the less fortunate—preferring to see the national system as a zero-sum game in which the economic goods are divided between haves and have-nots in unyielding ratios and percentages. The manufacture of revolution, or even the purging of older varieties of tyrannies, do not ensure nongenocidal outcomes. With the disintegration of Yugoslavia into at least six, perhaps eight, ethnic enclaves, one can witness how the overthrow of communism does not automatically usher into being a new set of democracies—more likely, a new set of racialist, ethnically grounded mini-states. When class, race, and religious stigmas are held as ineradicable sins, not subject to reform or repeal, the potential for genocide and mass murder is enormously increased.

The treatment of the Holocaust as a dialogue between God and Golem, as ineffable and unspeakable, serves to return the matter of death into the antinomic and Manichean tradition of original sin versus historical optimism. On the other hand, the treatment of genocide as an organizational problem for the United Nations makes of it a rather tepid affair involving the routine use of police to keep the offending state under control. Both approaches deny to this new "dismal science" of life-taking its full meaning and significance. It becomes the essential task of social research to move the analysis of life-and-death issues into the reality of nation-building, revolutionary regimes, and the authority of states as such. Such efforts are now under way and deserve support.[21]

If social science is to make its own serious contribution to holocaust studies it must get beyond the mystery of silence or the silence of mysteries. Still, however limited the theological and clinical analyses of collective death may be, they at least aim to spare mankind the repetition of some forms of genocide. To incorporate in the Jewish psyche the phrase "never again" requires an antecedent commitment to explain why genocide happened in the first place—and also why it may happen to other peoples "once again." Theologians must not presume an exclusive monopoly on meaning by insisting upon the mystery and irrationality of taking lives. The task social science takes upon itself remains in this area what psychoanalysis presumes to be its role with respect to neurotic individuals: a rationalization of irrationality. Only in this way can victory be denied to Golem and the struggle against evil be understood as a quintessential social task. This is a far greater challenge than standing in silent awe at tragedies that have befallen our tragic century.

11

Policy Research in a Post-Sociological Environment

The purpose of this chapter is to examine the present bifurcation of policy-making into domestic and foreign components and to urge a research effort aimed at unifying national policy by integrating its various components. Beyond this urging, my aim is to show that the act of making policy invariably involves decisions about events that take place both within and outside of a nation. This is not a claim for the superiority of any one segment of policy-making. What is important or trivial is determined within a means–ends continuum. In breaking down artificial barriers inherited from nationalist models, a policy process can be institutionalized that takes into account "shrinkage" of the world that is a direct consequence of new information technologies and, at the same time, incorporates the wisdom of classical social theories on the nature of power.

I have always had a hearty disdain for any article or book that begins with words like "toward an integrated view" of just about anything. This probably stems from a long-standing suspicion of general theories that seek to explain everything and, for the most part, end up by explaining nothing. I prefer the late Hans Reichenbach's tough-mined view of the world, as stated long ago in *The Rise of Scientific Philosophy*, in which explanation has meaning only to the degree that it entails the potential for prediction. But at this level of exactitude, social and political sciences have proven woefully inadequate. One need only look at major events from the 1989–91 revolutions in eastern Europe and Russia to the 1991 wars in the Middle East to understand as much. The utter failure of the social sciences to anticipate these events, much less predict the way in which such events would play out, should make each and every social scientist modest in the face of reality. Indeed, it is not simply the failure of analysis that is disturbing but, even more, the steady drumbeat of support for totalitarian regimes.

Beginning with a deflation of pretense and rhetoric is a risky opening gambit for one who urges a new style of policy-making. Yet, I dare to violate my own prohibitions in this instance because integration in the policy-making arena is precisely what is needed to help move politics and policy closer in a post–Cold War environment and to move explanation and prediction closer as well. The first step in this complex enterprise is to remove the debris of metaphysical dualism. We have taken far too casually as a canon the dualist premises of policy-making: one set of budgetary and allocation decisions for overseas needs and another for domestic consumption. But such a bifurcated vision of policy is transparently weak, polarizing communities into fratricidal interest groups. Only bureaucratic sloth keeps us from perceiving this.

We have all grown up with a social-science environment in which a dualism of foreign affairs and domestic affairs is taken for granted. Even the titles of major journals reflect a marked dualism. For example, we have *Foreign Affairs* and *Foreign Policy* on one side and *The Public Interest* and *The National Interest* on the other. As a result, and across the ideological spectrum, we take for granted this self-serving dualism, even though in actuality it is more honored in the breach than in the practice of policy-making. The purpose of these remarks is to reject this distinction and to urge that we reconstruct policy analysis and return the field to a unitary vision of political life, one that moves beyond decision making as something boxed into tiny windowless partitions that separate appointive from elective officials.

The literature on political economy has been compelled to consider conditions in which foreign and domestic policy intersect, as they do with discriminatory trade preferences based on global political coalitions[1] or the place of international pressure in explaining varying levels of protectionism.[2] However, there is remarkably little effort to derive from these specific examples a more general theoretical explanation of the present-day policy framework of advanced nations. The need for such a rethinking of basic premises is increasingly evident as we engage in our own "new thinking" on American national-security needs in an age in which old enemies dissolve and new ones arise with startling suddenness to replace them.[3] Just how such new frames of "multipolarity" are related to the "national interest" remains a central problem in need of solution.

Before pursuing this point, I want to emphasize that my position in no way denies national interest as an organizing premise either of state power or economic organization. To be sure, it seeks to reclaim expressions of the national interest, domestic no less than overseas, in the conduct of policy-making. The categorization of issues as domestic (like environmentalism) or of foreign-policy concern (such as the size or character of the armed forces) leads to a policy myopia of enormously dangerous potentials in which larger-scale national interests can be subverted in the name of one or another parochial movement. It is not my aim to deny the existence of multinational economies, regional alliances, or transnational pacts. Instead, I want to show how recent events have sharply curtailed the distinction between domestic policy and foreign policy; indeed, how they have increasingly come to intersect and thus change the face of theory and practice alike.

The domestic–foreign-policy dualism has been especially pronounced in, but by no means confined to, the United States. In some measure, this country, for two hundred years, has been preserved from the more severe consequences of warfare. The great exception, of course, was the American Civil War, which, as a result, continues to hold a special place in the consciousness of the people. Today, Americans tend to think of military conflict as an "overseas event," as if this pair of words somehow preserves the nation from the civil ravages of military struggles. Even in this age of nuclear energy and random terror, there remains a sense of vast oceanic borders protecting American shores. Given such an inheritance, it is little wonder that the force of theory separating foreign and domestic affairs remains quite intense and clearly intact—far more so, I venture to say, in the United States than in other parts of the world.

What finally seems to have snapped this dualism is the protracted military stalemate in the Middle East, which has remained since the early 1970s. I emphasize military stasis since the political positioning for the past two decades has been intense and anything but consensual. In a brilliant series of maneuvers, the Arab dominated Organization of Petroleum Exporting Countries (OPEC) compelled a reconsideration of American foreign policy through its oil embargo. Although this embargo was relatively brief, incomplete, and revealed many fissures within the Arab Middle East, its consequences on the American public proved to be enormous and long lasting.

In a near panic state, American automobile drivers experienced substantial delays at the gasoline pumps, faced inflationary run-ups in gasoline prices, and were compelled to change their personal and consumer behavior. Thus, events in the Middle East, while geographically remote from the American mainstream, had an immediate and lasting impact on American behavior. Attitudes toward Israel slowly shifted to neutral, however hesitatingly, while technological pressures to develop electric-powered vehicles and alternative sources of heating (still largely in the planning rather than mass-production stage) became part of the rhetoric if not the reality of the domestic landscape. Above all, a strong sense of American dependency rather than American domination emerged.

While this lessening of economic autonomy was taking place, quixotically, a noisy band of social scientists was beating the drums for a theory of dependency. They dusted off Leninism to prove that the foundation of the world economy was rooted in Washington and Wall Street decision making and that the rest of the world, especially the less-developed nations, were beholden to the American colossus. Rarely in the annals of scholarship has the gap between the real world and the academic world been greater or, ultimately, more transparent. For with the OPEC-inspired oil boycott, an entirely new set of relationships emerged. A huge transfer in real wealth took place, in which the advanced nations were able to maintain a modicum of stability in trade balances by selling huge amounts of military know-how in exchange for black gold. One can detect in everything from the bunkers of Baghdad to the Scud missiles of the Iraqi army the consequences of such huge fiscal transfers. As part of any serious reconstruction of policy theory, we should also brush away this remnant of dependency theory that continues to litter the academic landscape, real-world events notwithstanding.

The OPEC embargo created a shift in American policy-making, both in domestic and foreign affairs. The Alaskan and Gulfport oil deposits were developed. Arrangements were made for better cooperation with American allies, especially with regard to the off-shore deposits in the North Sea. Front-line nations like Saudi Arabia came under the American military umbrella. The United States performed an honest-broker role in resolving Egyptian–Israeli border disputes. Such measures changed domestic energy equations no less than overseas military capabilities.

There also took place a variety of subtle shifts of a seemingly more parochial character. Legislation encouraged development of clean-performance engines rather than gas-guzzlers; surcharges and taxes were imposed on engines with low milage per gallon; family-owned gas stations declined; and the marketplace was increasingly dominated by the large oil-refining companies. Consumption of fuels did, indeed, decline, but in a larger sense, the petroleum crisis was resolved as it had begun: less by dedicated policy than by a series of changes in the way nations created stable bilateral business relations with one another. A delicate global infrastructure emerged to fine tune issues of policy and economy alike. The circuitous route of policy should be understood in terms of these large-scale equations.

When bilateral and regional alliances broke down, as they did in the summer of 1990, the crisis that was capped by the invasion of Kuwait by Iraq, the concerns quickly spread to the control of oil reserves. Parallel issues emerged, such as the connection of Saudi Arabia with the United States, the proximity of Jordan to Israel, and the norms governing relations within the Arab League politically and OPEC economically. Driving these foreign-policy shifts was an awareness that nothing short of domestic survival was at stake. Norman Podhoretz is quite correct to say bluntly that "Oil *is* the life-blood of our civilization and any threat to the supply of it *is* a threat to our way of life." This view is curiously echoed from a distant ideological position by Charles William Maynes, who similarly declared the conflict in the Gulf as "The War for Oil."[4] Whether this be termed a "vital interest," a "national interest," or a "foreign-policy concern" is pure semantics. What emerges clearly in the current Middle East struggles is the unitary character of policy as a response to the singular character of raw-material needs.

Within hours of the invasion of Kuwait by Iraq, the following domestic events took place in the United States: the price of crude oil rose 10 percent on the New York Mercantile Exchange; the Dow Jones average dropped more than 50 points; the price of United States Treasury bonds fell dramatically; interest rates rose on the threat of a new inflationary spiral; and airlines added a surcharge for all domestic flights. By the same token, and on the same exact scale, when hostilities broke out between U.S.- and British-led U.N. forces and the Iraqi military, all of these same economic indicators reversed their patterns: the Dow Jones average increased by more than 100 points; the price of United States Treasury bonds rose; and interest rates came tumbling down. Airline prices persisted, but this was a function of armed conflict and the fear of terrorism rather than of an oil shortfall. Indeed, there was as much talk of an oil glut as of an oil shortage, with a corresponding sharp reduction in prices.

Similar events occurred in other industrialized nations of the West. But the principle was consistent: a major international crisis precipitated major domestic crises. Hence, policy formation was interlinked. From the use of military means to instill a blockade to the development of a crash program for producing all-electric automobiles, old dualisms were based on elongated time and space. The distance between nations meant that it might take days or months for actions taken to affect events elsewhere. But today, military decisions about events taking place in remote geographic areas are measured in terms of seconds, minutes, and hours; thus changing both the context and content of policy-making.

If the 1970s were dominated by the petroleum and energy crisis, then the 1980s could be considered the decade of drug-related crises. Again, this is an area in which the thin line between foreign and domestic policies was so often traversed that the very distinction itself finally had to yield to a higher truth: that events in Marseilles or in Bogotá were directly related to the price and supply lines of drugs in Miami and Chicago. Indeed, relations among drug suppliers, distributors, manufacturers, and users became part of a new mosaic—defined as illegal and, hence, played out on a cops-and-robbers stage; nonetheless, global and local in every way.

In meetings between leaders of Andean cocaine-producing nations and North American consuming nations, the big issue is where to place prime responsibility: on the streets of big cities where drug use is rampant, or in the poppy fields of hamlets and villages in Peru, Colombia, and Bolivia, where cocaine is grown and harvested. This is hardly a moral discourse, since at stake is a coordinated effort to eradicate drug use while not upsetting the economic stability of these South American nations. To be sure, the availability of a product or a crop—whether it be petroleum or drugs—is key to market prices as well as political attitudes. A key element in the collapse of the distinction between foreign and domestic policy is the need to create scarce supplies in the face of heavy demands. Once again, the United States found itself the dependent victim rather than the dependency creator.

The drug problem highlighted the raucousness of much of what passes for sociological wisdom in this era. The fashionable belief that drugs were little more serious than cigarettes or whiskey went up in the smoke of extraordinary statistics showing the high correlations of crack usage with arrest and conviction rates for serious crimes in U.S. cities. The fashionable middle-class notions of drugs as a "recreational" phenomenon characteristic of an affluent society also collapsed as occasional usage spilled over into spiraling rates of addiction among American youth. In the United States in 1990, according to the National Institute for Drug Abuse, twenty-seven million Americans, or 11 percent of the population, used some form of illicit drugs at least one time during the year. And this is an area of acknowledged underreporting. And even if drug usage and drug addiction are by no means identical, the numbers are sufficiently staggering as to cause all but the most confirmed advocates of the drug subculture to pause.

The notion of drug purchases in nonpunishable zones of sale and distribution, once so widely touted as a livable "Dutch" solution, floundered on the hard realities. It was obvious that such a policy proposal, whether in the name of legalization or *laissez-faire*, only served to sanction a two-tiered society of drug users and was not a broad social solution to drug abuse as such. The most liberal theorists on drug decriminalization became trapped in a quagmire of antidemocratic political theory—in which permissiveness was transformed into a theory of tolerable levels of human waste. Such theories were proffered by the same people who attacked defenders of military solutions for offering similar approaches to acceptable levels of human waste in war zones.

One might well argue that, *de facto* if not *de jure*, using drugs and selling drugs have become nonpunishable crimes. Thus, even though these are crimes with victims, they have not been punishable (at retail levels at least) to a very large extent. What happens is a crisis in advanced societies that cannot be mandated on policy grounds—except by raising the stakes: permitting higher prices for poorer drug services and goods and higher risks in drug purchasing and acquisition. As part of this trade-off, the organized worldwide drug network settles back into a basic business equilibrium. Not unlike for oil, prices become rationalized, goods and services flow in relatively uninterrupted fashion, and a new class of entrepreneurs emerges and changes the character of global relations, without necessarily shaking the foundations of the social order as such. In short, the explosion of drugs, from crop management to street transfers, involves many nations and people.

A third example of the breakdown of domestic–foreign-policy distinction is the disease which has come to be known as AIDS. Presumably originally from Africa, it is now a plague of worldwide proportions transmitted by tourists, soldiers, and assorted travelers from big chunks of the less-developed world. This now third most disastrous cause of death was virtually unknown only a decade ago; today, the budget for AIDS research in the United States exceeds that of cancer research. Again, we see a total reintegration of policy as a result of a network of events that respected few, if any, national borders.

Not only the AIDS virus but also its treatment policies are universal. In the specialized agencies of the World Health Organization, the only administrative bodies within the United Nations that truly function in an ongoing capacity, we find policy being set for a variety of chores: research as such, relevant drug innovations and intakes, and pooling of medical results. The very hint of possible contagion or plague is often sufficient to mobilize otherwise secretive and recalcitrant leaders into some form of policy-making. Dictatorships, dynasties, and democracies alike are involved in pooling information that is central to policy as such.

One may also view the globalization of policy, and the breakdown of the distinction between domestic and foreign policy in a quite different way: from the top down as it were. New abilities to explore outer space have also reinforced our awareness that the Earth is one planet—flanked by Mars and Jupiter on one side, and by Venus and Mercury on the other. However trite such statements appear, in this celestial family, the Earth counts as one. As a result, policy involves a variety of environmental considerations not confined to any one nation. While it would be naive and utopian to suggest that space exploration in itself will lead to universal brotherhood and a collapse of rivalries, it does reduce the relative weight any serious person can give to the interests of the particular nation-state.

Even if one is uncomfortable with such astronomical levels of policy discourse, and I count myself among that number, even confining policy concerns to earthly matters must give us pause. Petroleum spills, whether accidental (as in Alaska or the Gulf of Mexico) or intentional (as in the Persian Gulf), affect the vital interests of several nations. Fishing policies of one nation affect the notion of sovereignty on the part of other nations, as witnessed by bitter arguments over the two-hundred-mile zone versus the thirteen-mile restrictions on off-shore rights. Radio and television stations in one country relay signals to another country—possibly with a quite different political system. All of this is well known, of course, but our political scientists and sociologists continue to act as if such events are somehow irrelevant because such concerns have been either designated as part of national policy, or seen as parochial, having had a minor impact in the past.

Asserting a dualism between domestic and foreign policy is clearly untenable in any of the aforementioned contexts. Why then do we continue to hold firm to fixed stars that are long dead? I suspect that the answer lies in the appeals to tradition at one level and to law at another. I mean by "law" the ability of the nation-state to continue imposing sanctions, restraints, issue edicts, and regulate the behavior of its citizens. The Age of Aquarius has not dawned. The nation-state remains very much a force to be reckoned with. To be sure, events in the Middle East and central Europe indicate that priority of national interests rather than a new world order would be the more probable outcome of immediate struggles. National policy-making will not dissolve in one fell swoop.

It would confuse realms of reality to assume that the existence of the national state, even the national interest, confirms the dualism between domestic and foreign policy. For in an environment in which policy-making is seen as a unitary rather than binary phenomenon, the nation-state will remain very much a force. Perhaps it will be even a greater force than it is at present, since it is precisely the legal and military capacities of a nation that alone can impose that nation's will on all citizens and force negotiation with other states having their own particular sets of national interests. In other words, any sense of a global environment is the sum of its national parts. The moral force of international organizations remains, for better or worse, a function of the military force of national systems.

This point is dramatically illustrated by the Comptroller General's 1990 Annual Report of the United States General Accounting Office,[5] perhaps the finest social-scientific establishment in the nation. The report lists as highlights its activities in research projects with policy implications in the following areas: military presence of U.S. personnel in NATO Europe; black-marketing of U.S. goods in South Korea; training standards and aviation security at foreign airports for U.S. personnel; immigration reform and employer sanctions; U.S. weapons and technology threats; foreign aid and efforts to improve Latin American judicial systems; the dichotomy between U.S. export policy and antismoking initiatives; electronic warfare and air-force jammer programs; and global airline competition and marketing practices. One can continue this list, but the point is clear enough. What currently drives policy research is a mix of national, regional, and global concerns. We find the same set of considerations present in social science research. For example, the report of the

Research Triangle Institute in North Carolina[6] is organized in terms of program evaluation research, population and family studies, alcohol and drug abuse, criminal behavior, mental health problems, hazardous-waste management policy, natural-resource damage assessment, risk management, energy resource planning, and market assessment and forecasting. Again, one is struck by the replacement of the traditional five fields of social research (psychology, sociology, economics, political science, and anthropology) with an entirely new range of functional categories.

It is fair, then, to say that we are at the dawn of a new era of unitary policy-making; and we need to reconsider organizational frameworks inherited from a past in which the dichotomy of the national and the international still made some sense. Clearly, the velocity of change, the shrinkage in measuring distances between places, and the shared physical as well as social environment make all past paradigms obsolete. The realities of social research and policy have already transcended old boundaries; only the forms remain to be changed.

This argument for a global context of the national interest must be made unambiguously and without the usual rhetoric of one-worldism, for otherwise, it can be misconstrued as an appeal for a return to world federalism or as the recently declared search for a new world order. International considerations that do violence to the national interest have proven bankrupt at the empirical, as well as at the ethical, level.

The nation remains the repository of policy-making at more vigorous levels than in the past, if only as an organizational vehicle for executing a variety of tasks common to all. In this sense, the modern nation is a mechanism for integrating the complex and specialized needs of its citizenry. The creation and distribution of essential goods and services remain in the province of the nation-state. The policy apparatus provides a body of knowledge, language, and culture as part of this process. But policymakers in a democracy are still responsible to the people whom they presumably serve. The policy apparatus integrates the mix of considerations in the provision of goods and services, ensuring that this is done with intelligence and skill. For want of a better term, this is the essential "field of battle" on which policy agencies operate.

As long as individuals perceive their own interests in discrete utilitarian terms, parochial politics and policy-making will be with us. The breakdown of the distinction between foreign and domestic policy-making does not signal an increase in the spread of universal moral properties of goodness and righteousness. It does signify some coming to terms with the need to fashion policies that get beyond the tug of narrow interest-group demands at one end and the pull of traditional isolationist pressures at the other.

One hopes that it is still possible to move beyond the lethal politics of what is euphemistically called "pluralism." The constant drumbeat of demands without any parallel recognition of obligations leads to a policy-making context in which domestic requirements are seen not only as different from foreign policy requirements but also as part of a contest between interest groups. The key philosophical obstacle to a unified theory of policy-making is the assumption that an economy is a fixed pie, and therefore one must fight for a maximum share of that pie. As a result, the notion of economic growth no less than political common good is sacrificed on the altar of a dangerous and narrow utilitarianism in which the whole is less than the sum of the parts. All that is left of the American nation are the discrete parts.

The need for an integrated vision in policy-making is at rock bottom a reassertion of political values that have guided the American nation from the outset.[7] Administrative differentiation has disguised valuational breakdown. The various calls for streamlining and reorganizing government must be seen in the broader context of organizational malaise that threatens to strangle policy-making as such. Once the administrative and ideological obstacles to a unified approach to policy-making are identified, we can move ahead with the task of reconsidering national priorities in the context of global constraints and possibilities.

In identifying at least a few of the obstacles and issues facing the reconstruction of policy-making in America, I hope that I have brought to greater attention the interconnection of all phases in the policy apparatus. In doing so, I have also tried to make a case for thinking of policy-making as a unitary phenomenon and not as a cacophonous cluster of domestic and foreign noises with differentiated agencies to match. A recent work by an environmentalist puts matters succinctly: "People in the modern, affluent democracies must demonstrate awareness that their standards of living, or those of their children, are directly threatened by what happens to the tropical forests, the ozone layer, Lake Como and the Finger Lakes, and the birds of Antarctica. They must acknowledge that what happens is irrevocably linked to their energy consumption habits."[8]

This is quite true. But it is also time for environmentalists and special-interest advocates of all sorts to face the fact that the activities of a drug cartel in Medellin, Colombia, a military adventure in Kuwait, or a crackdown on separatist movements in the Baltic states are no less linked to energy consumption habits. To continue on our present course is to make the policy-making apparatus hostage to the very parochial interest groups that a truly national policy is dedicated to overcome.[9] To embark upon a new course of action, which ironically entails a return to basic value premises of the democratic system, will once again permit policy-making as a whole to serve the interests of the American people as a whole.

What I have in mind by increasing the democratic character is simply to bring policy-making with its appointive and expert features closer together in function and structure to the American elective and populist traditions. The fragmentation of society into polarized interest groups, the bifurcation of the world into the domestic and the foreign, and the dominance of technocratic modes of policy-making have wrecked havoc on integration in decision making no less than in policy formation. Interestingly, the same sorts of concerns are now expressed as essential causal agents in the stagnation and backwardness of such potent Third World nations as Argentina[10] and Mexico.[11] As a result, this proposed search for an integrated approach to policy-making is hardly confined to the United States. One hopes that an essential by-product of the approach herein taken will be a noticeable enhancement of responsibility at the top no less than broader participation in the political process by those who now see policy research as remote and a barely tolerable necessary evil. Of course, such a conclusion can be construed as a wish-fulfillment approach to policy-making. But then again the search for linkage and fusion of the foreign and the domestic might well be subject to the same charge of utopianism.

The paradox of policy is that in order to get beyond polarity one must get beyond policy as such. That is to say, to arrive at a unitary vision of policy we need a unity view of moral purpose. With all due regard for criticisms of the Kantian *Grundlegung*, it remains the essential guidepost in any serious effort at integration. It is worth citing at some length.

> This principle of humanity and every rational being in general *as an end in itself* is not derived from experience, first of all because of its universality, as it extends to all rational beings in general, and no experience could be sufficient to determine something about that; secondly, because humanity is here represented not as an end for man (subjectively), that is as an object that one really adopts as an end for oneself, but as an

objective end that, whatever ends we may choose to adopt, is supposed to constitute as a law the supreme limiting condition of all subjective ends, thus must originate from pure reason.[12]

With the virtual collapse of Marxism–Leninism as a codebook of policy management based on a presumed "correlation of forces," we can now move beyond pure interest-group theory. In moving from the particular to the general, we also get beyond specific variables of nation, class, gender, and community—namely, factors which stimulate, if not create, cleavages and conflicts—to a "human" point of view. This is not to deny the existence of classical factors that continue to reemerge as a backdrop to the century. It is to deny that variable analysis alleviates the need for ideals and goals in human affairs.

The dualism between the national and the international came about in an environment in which defense of the national realm became the touchstone for foreign policy. In that sense, the Hegelian worldview found in *The Philosophy of Right and Law* aimed to displace Kantianism. For Hegel, there really was no general will, only a series of particular wills—and these could only be coerced into unity through the Will of the State. There was no acceptable international view, only a national view. Hence, in this way, dualism was built into the structure of both nineteenth-century Hegelianism and twentieth-century Marxism.

The evolution of a consensual foundation of democracy was held at bay during the past two centuries, in a series of life-and-death struggles against the conflictual foundations of totalitarianism of all sorts. We began to view the dualistic nature of policy as the only realistic, or at the least non-suicidal, way the democratic West could cope with antidemocratic systems and ideologies. And while the collapse of eastern European and Soviet communism is by no means the end of totalitarian regimes—the survival capacities of China, Cuba, and North Korea attest to that—one hopes the terms of the dialogue in the period ahead can return to and amplify a Kantian vision, an integrated view of values, and, hence, a unified vision of policy-making.

Admittedly, this is a highly speculative but one hopes not a spurious outcome to an examination of policy-making in a post–Cold War environment. The machinery of such an integrated policy-making environment need not be global in every instance. It can be regional, such as the European Economic Community, Organization of Petroleum Exporting Countries, Organization of African Unity, or even Interpol. It might be objected that such regionalist views disguise higher and more advanced forms of Hegelian separatism. And in truth, there is such a tendency in a regional, in contrast to a global, organization; the United Nations being a case in point.

However, the habit of cooperation between and among nations is an essential countermarch to purely nationalist striving. It also moves counter to the process of Balkanization that is nowhere better exemplified at the moment than in the Balkans. The movement beyond duality, beyond policy as either national or global, can succeed; a regionalism free of chauvinism disguised as hard-boiled self-interest can provide the answers. Admittedly, this is a tall order. However, by moving the question of policy to a philosophical and ethical agenda, we hold out the distinct possibility of moving beyond polarity as such.

12

Prediction and Paradox in Society

The identification of social science with social advocacy has reached such pandemic proportions in American sociology that it is time, indeed the time is long overdue, to step back from the precipice of partisanship if the worth of serious analysis is itself to be preserved. Instead of being a possible consequence of decent social research, advocacy has become the very cause of social research. We have taken the chief weakness in the structure of knowledge about society (namely, the propensity to ideological thinking) and turned it into a first principle of the research process.

In attempting a reconstruction of the social sciences, a return to first principles is necessary. But just what are these "principles"? Where does one locate them and how does one avoid the same dismal condition in which we now find ourselves? I would suggest that the answers inhere in everyday life. That life is often well described by journalists and storytellers, for whom the heavy weight of social reform does not exist; or at least is sufficiently insignificant so as to permit the examination of human behavior as a story worth telling in its own right—without self-righteous presumptions as to what should be the proper goals of such behavior. Answers are also to be found in the legal codes, those aspects of jurisprudence which recognize that conflict of interests are authentic and valid and are, hence, subject to fine tuning and adjudication.

From Empirical Prediction to Moral Paradox

There is a growing recognition that there are serious limits to treating social research as a simple extension of physics research.[1] There is, therefore, far less attention paid to what is the basis of social research. To think about this with precision involves a return to the sociology of everyday life, for everyday life is the location of the social-scientific vocation. The social situations in which ordinary people find themselves are filled with paradox and with individually distinctive and often contradictory views and attitudes toward the common objects and experiences which make up ordinary life.

There is a somewhat tattered, but not quite broken tradition in sociology to take seriously the propositions of everyday life as reported in and expressed by newspapers and popular magazines. To be sure, it is a tradition that is mostly ignored or, worse, openly frowned upon, as an inheritance that moves counter to professionalization. I suggest that we return to the sociology of everyday life, and away from its pathologies, to arrive at some sort of revitalization of the search for social knowledge.

Everyday life does not dissolve under the timorous assaults of medieval craftsmen, nor have the popular media become obsolete. Indeed, one might well claim, as Dorothy Nelkin has, that such media increasingly have become responsive to the findings and messages of social science—as expressed by high, if selective, reporting of major findings of public issues.

One reason for the disuse of newspapers may well be the specificity and topicality of good reporting. But such features can also be employed to gain a measure of sociological insight. Thus, Robert E. Park's work on the urban life cycle was profoundly informed by the "big news" of the big press of Chicago.[2] Robert and Helen Lynd's work on Middletown (or Muncie, Indiana) was clearly done at a time when he was a committed journalist in the early 1920s.[3] And Elwin H. Powell's studies of suicide in Tulsa and Buffalo derived from a careful, systematic review of vast newspaper archives.[4] My own review of sources will be far more modest but I trust no less convincing.

The suspicion remains that such use of public materials in the fashioning of public events is severely limited in its value to the social scientist. The purpose of this chapter is to address these concerns—specifically, to show how the world of the newspaper can not only inform the empirical researchers in social science but also assist us in developing a new theoretical foundation for sociology as such.

Admittedly, this is an ambitious undertaking in a limited context. Still, to the degree that newspapers, far more than the purely visual media, continue to inform people of the nature of their societies, the task is worth pursuing—if only for the limited purpose of displaying the ways in which new theory construction in social research can take place outside the cocoon of inherited doctrines and methods.

My basic thesis is that the world of society is one in which conflicts of value take place and that these value differentiations are not matters of right or wrong. Those few absolutes that do exist are readily enough disposed of by courts of law, if not outright and direct retaliation. Rather, the major conflicts present themselves to us as paradoxes, or moral puzzles if you will. These are sometimes resolved by the remedies supplied by courts of law; sometimes by force of arms; and yet at still other times, they are not so much resolved as dissolved by changing circumstances.

The function of sociological theory, in contrast to ideological posturing, is to put into full view the nature of the specific paradox that divides people along class, religious, gender, racial, and ethnic lines. This is not to celebrate conflict, but rather to show the forces at work in moving a specific paradox to resolution or dissolution. In this way, the same sort of analysis that characterizes the study of foreign affairs, with the same sort of clinical detachment when done properly, can make its way into the study of domestic issues—or at least issues with a more obvious national or social setting.

In a sense, I want to continue the oft-interrupted march "out of utopia" offered by Ralf Dahrendorf in the 1950s—that is, to identify the paradoxical nature of ambitions, drives, and motives fueled by the conventional stratification. We must finally get beyond the notion that a social variable is a political cause. Even worse is the presumption that political causes are the true vocation of the social scientist.

The assertion of class conflict is not an *a priori* demonstration of the virtues of proletarian manhood. And the existence of sexual differences in the marketplace is not prima facie proof that the proletarian womanhood requires a party unto itself. In the real world (or better, the world as described in daily newspapers), we come upon a constant stream of painful choices made by good people in difficult circumstances. With this I conclude my prologue and enter into the main thrust of my argument.

Paradox is the essence of the social-scientific standpoint. It is the only way we can appreciate the full range of motives and goals of everyday life, and do so without recourse to dogmatism and a presumption that the essence of science is having answers for people instead of listening to them. By no means is acceptance of paradox a surrender to confusion or subjectivism. Rather, identifying paradox in the nature of social objects is to return social research to its moral sources and, hence, to a greater awareness of the anguishing difficulty in viewing the social order and the human subject with sensitivity. All this is to say that there can be no good society without good people. Thus, the task is to locate the wellsprings of moral order in human conscience.

It will be readily perceived that the foundation of a social science is quite distinct from that of a physical or a biological science. For to consider paradox as a "center of gravity" is at the same time to deny that centrality to the notion of atomic prediction. It must be said in all frankness that the aping of the physical-science model has cost social research a ton of false starts and painfully wrong outcomes. For one must start with the person, the individual, as the setting of human interaction, and not with models inherited from other disciplines that feature the study of inanimate objects or reflexive or reactive members of the animal kingdom.

I do not wish to deny the place of prediction, when possible, nor, for that matter, do I intend, by an emphasis on paradox, to lose the search for objectivity in analysis and solutions in problem solving. To recognize the place of paradox in social settings is not to open a Pandora's box to a pure relativism, or an empty-headed subjectivism. Quite the contrary, paradox is rooted in objective circumstances, in the very nature of human nature. That is why the study of social systems is uniquely different from the examination of physical space.

The changing nature of the circumstances or the situations in which people find themselves is also the primary stimulant to doubt inherited formulas and models. The beauty of theory is the explanation of specific events; the curse of theory is the use of the overgeneralized same explanations to interpret unlike events. In this sense, the key to a reconstructed social science is not the size or importance of the subject under discussion but the generalizability of the finding. One can study the behavior of two people on a minisubmarine and arrive at some important staements of human behavior, but can one apply these ideas to the behavior of superpowers without spouting absolute rubbish?

The prideful belief that sociology is more important than psychology because the size of the "sample" is greater must be rejected. At the same time, the belief that larger samples yield greater truths is no less suspect. The science of genetics came about through the study of characteristics of peas. The ramifications of that science did not depend on huge chunks of the world but upon the insights possessed by a single individual into a tiny portion of that world, insights that had profound consequences.

The social sciences, still mired in empty-headed discourse about "micro" and "macro" ranges of analysis, run the risk of thinking that big is beautiful or small is perfect. If the quality of research were solely dependent on the size or the frame of analysis undertaken, we would witness a rush to that particular size or range. But while such considerations are not entirely irrelevant in establishing evidence, neither are they especially illuminating. It is better to examine those elements at work regardless of the size of the object being studied. With that I return to the sociology of everyday life, the social paradox, and the moral dilemma.

Paradox and Pain

Let us then proceed to look at a sample of anguishing paradoxes reported within a one-year time frame. Doubtless, these can be multiplied a hundredfold. But our purpose is to illustrate in the broadest possible way the reason why prediction, in the specialized sense used in the physical sciences, must yield to paradox.

My first illustration is rather typical. It involves a public-interest group, aiming to do good, the efforts of which ended in the death of the very person it sought to assist. The Appalachian Research Fund, a public advocacy group, successfully argued, in a West Virginia lawsuit, that psychiatric patients should be given the freedom to live with as few restrictions as possible. The woman on whose behalf this was filed, Ella Hartley (indeed, it was known as the Hartley decision), was found drowned after wandering off during a visit to the doctor.[5]

One might argue that this is only one person who died in consequence of this liberal ruling; or that the mental-patient population at Huntington State Hospital fell from fifteen hundred to ninety. One might argue the opposite: that hospitalization is precisely what is supposed to prevent such catastrophes. It might also be noted that the size of the homeless population rises proportionally to the decrease in hospitalization for mental causes. Hence, the tax burden is merely shifted from one agency to another rather than diminished. Clearly, asking the question of right or wrong in this situation is simply meaningless.

A second illustration is similar in type but not in consequence. A fifteen-year-old girl, the daughter of Elizer and Maria Marrero, was chained to an iron pipe at home. This confinement lasted for roughly three months. She was allowed to eat and watch television but had to use a bucket for her bathroom needs. When their actions were discovered and they were arraigned, the parents argued that this was their last resort in attempting to prevent their daughter from being caught up in a network of heavy drug use, from being away from home for weeks on end, and from possible rape. When queried, the daughter confirmed the truth of the allegations and stated that she entirely approved of the remedy taken by her parents, pronouncing her love of them.[6]

The Child Welfare Administration of New York had no choice but to reveal and intervene in this situation. But when one considers the parents' absolute desperation over the personal risk to their teenage daughter, was it hatred or love that prompted such a Draconian measure? In addition, for what length of time can an association of child welfare assume a custodial role, and to what advantage? It might be argued that such an example converts a private trouble into a social problem. But such abstractions count for little in the immediacy of a problem. The paradox is only enhanced by the fact that the daughter accepted and even welcomed the physical restraint on her movement. Thus, our paradox does not even afford the usual comfort of the claims of child welfare against parental abuse.

This brings us to another variation on the theme of child abuse: photographing children and their families in the nude. Jock Sturges, a San Francisco photographer, has taken a wide-ranging series of photographs in black and white, and later in color, of children and their parents. These photos were taken with the permission of the families involved, but Mr. Joseph Semien, the film processor for these images, was arrested under California State law on felony and misdemeanor charges. These photos were often commissioned by "naturists" or "nudists" to illustrate their views as well as their bodies.[7]

To what extent are the photos of Sturges different from those of Mapplethorpe. If pornography is to be defined in terms of the images alone, should not punishment be similar? How does one define obscenity or the intent to be obscene? Is the display of genitalia to be construed as obscenity? Is nudity as such to be uniformly prohibited? Do nudists, seeking to propagate a viewpoint, have the right to introduce children under the age of twelve into such publicly sold and distributed photos? Does violation of the law come in the processing of the film or the sale of the prints? It is not the point of this exercise to resolve all legal situations, but to highlight the wide gulf between those who see such photos as freedom of expression and those who see them as a violation of laws pertaining to child pornography.

Another paradox that sits at the legal intersection of national debate about freedom of artistic expression and the use of public facilities for religious activities is Craig Martin's mural of a crucifixion scene which dominates a panel of the auditorium of the Central School in Schuylerville, New York. One group wishes to see the offending mural removed because children will experience extreme discomfort and anxiety at what is perceived to be a portrayal of the crucifixion of Christ in a public, secular school. On the other hand, other affidavits collected argue that the mural indicates simply "man's inhumanity to man." It was further argued that the mural was intended as opposition to the Vietnam war, which was raging at the time the painting was done.[8]

The opponents cite the "wall" separating church and state and claim that such a depiction of the crucifixion of Christ, and its very size and placement, creates in the minds of young people a misconception that the school endorses Christianity as a state religion. Supporters argue that this is no more offensive than a mural of the American flag or a painting of the Revolutionary War. The Board of Education in the school district plus an overwhelming number of community residents wish to see the painting remain. What role should community sentiments play in decisions of this nature? And do they supersede federal decisions on religion and the state? Once again, there is a perceptual problem: is the depiction of a crucifixion necessarily Christian? Whatever the court rulings in such matters, they appear again and again—indicating that paradox involves moral principles and not just perceptions.

Battles over the symbolism of the cross and flag haunt the political process directly. At state and federal levels, prohibitions against flag burning have been struck down. Yet, the issue reappears in a thousand guises. In one of the more recent examples, the Judiciary Committee of the U.S. Senate fashioned a flag protection act to blunt momentum for a Constitutional amendment. The argument runs that a blanket ban on defacing the flag, without regard to the perpetrator's political message and motive, serves the purpose of preventing undue assaults on civil liberties. The opponents of such legislation argue that it would embrace the symbolism of the flag and is akin to burning a village in order to save it. In effect, the argument holds that prosecution for flag burning violates the very same First Amendment such advocates seek to defend—whether by Constitutional amendment or congressional action.[9]

The ultimate paradox is the nature of the flag itself. What is "the flag," and what does its willful destruction signify? Is it a logical impossibility for a person who burns a flag, one of many millions that exist, to injure the symbolic meaning of the flag as such? In what sense are reality and symbol inextricably intertwined? The situation also involves choosing between Constitutional preferences and political instincts. Since the Bill of Rights leans heavily toward protecting individual acts, even unpopular ones, while the overwhelming majority of the American people are repelled by such acts of vandalism, does a piece of legislation or an amendment aim to satisfy public appetites at the expense of minority rights? And where will this road lead if followed? On one hand, there is the potential harm to system legitimation in flag burning; on the other, there is the potential for system derailment in banning what are, after all, random and occasional acts.

The problem of preserving civil liberties becomes far more of a challenge in the announced "war on drugs" than in flag-burning incidents, for involved is a massive, federally supported effort ranging from a proliferation of urine testing in the work place to police roadblocks of suspicious neighborhoods and even to waging mini warfare to capture foreign rulers (such as Manuel Noriega) held to be responsible conduits of the drug trade. Indeed, the Supreme Court has held that the Constitutional prohibition against unreasonable search does not apply to American law-enforcement agents who seize property of foreigners in foreign lands.[10]

The concerns raised are social as well as legal. To start with, such search-and-seizure procedures seem to take place disproportionately more in poor and black neighborhoods and in public-housing facilities than is legally tolerable. Beyond that, there is a charge that law-enforcement agents cut procedural corners to what amounts to kangaroo courts aimed at expelling suspected drug dealers. And finally, there is a contagion effect, a widespread belief that even if civil rights are trimmed (as in random rulings involving the drug testing of public employees[11]), the magnitude of the drug epidemic warrants such curbs.

Establishing policies in this area thus depends on far more than empirical or statistical information on the magnitude of illegal drug use and trafficking. It involves nothing less than the price a society is willing to pay in ordinary protections of property and life to reduce such drug use. For example, recent court rulings have affected the Sixth Amendment right to legal representation, others have involved Fifth Amendment protections against self-incrimination by demanding from defense lawyers the names of people giving them in excess of a certain amount of money. At what point does a society, to protect its very survival, curb rights guaranteed by social contract? Clearly, the drug arena is but one of many where such matters are fought out. But once again, it must be seen that the essence of the analysis is the situational paradox, not the fatuous policy declaration.

The situational paradox becomes far more complex when groups of individuals who perceive themselves as deprived citizens are involved. One such area is housing, as it affects children and the elderly. Retirement communities of the elderly are clearly disturbed by new rulings that prohibit discrimination against families with children. Such legislative and legal relief is aimed at protecting the rights of young families to move into complexes geared to the needs of children. Ugly generational conflicts that have arisen over issues like Medicare now have moved into housing. And even though specific exemptions for the elderly are often made, such exemptions are so vague that many of the communities involved are fearful of risking heavy penalties and open their doors with reluctance to families with children.[12]

Interest groups ranging from condominium associations to the Senior Civil Liberties Association argue that fair housing legislation violates the right to privacy and takes property without compensation. The social components of this are the fear on the part of the elderly of increasing stress and turmoil and a breakdown of the ability of such facilities to address the special medical and transportation needs of the elderly. In contrast, there is a growing recognition that age segregation is just that—segregation.

The right to choose which age group one wants to live with comes up against the right of individuals to choose what neighborhoods they wish to live in if they have the means to pay. The issue becomes the political potency of the elderly versus the economic purse strings of the younger who are still in the work place. Here we see one of many areas in which large social groups defined as good are pitted against each other. Paradox turns into anguish. "Allies" in macroscopic causes readily become "enemies" in the intimacies of interest-group struggles.

The closer we examine the nature of paradox, the nearer we are to a sense of the social world as a place in which rival "rights" are involved. It is not the biblical struggle of right versus wrong, but the pragmatic conflict of groups involved in struggles for scarce resources that impress. In a recent article on Detroit's racial struggles, the Israeli writer Ze'ev Chafets captures this terrible sense of competing rights.[13]

Tom DeLisle, a former white alderman in Detroit, summarizes the situation thus: "In metropolitan Detroit today, fear is the most pervasive single factor. . . . Everything goes back to the racial situation. Detroit has been the first major American city to cope with going from white to black. And whites left. That's the American way. People have a right to move in, or move out. Who wants to put their kids in a situation where they are likely to be crime victims? That's as basic as life gets."

Arthur Johnson, one of Detroit's four police commissioners responds with defiance and vehemence: "Detroit has unjustly come to represent the worst in America. If they make that stick, it is possible to justify our neglect and separation. Whites don't know a god-damned thing about what's gone wrong here. They say, 'Detroit had this, Detroit had that.' But economic power is still in the hands of whites. It's apartheid. They rape the city, and then they come and say, 'Look what these niggers did to the city,' as if they were guilt-less."

These are not exactly sophisticated sentiments, nor are they especially unique. Indeed, they are so commonplace that it is time to realize that the manufacture of policy, whether to isolate and cordon a city or pump great amounts of fresh capital into a city, does not derive from empirical analysis but from the valuational base on which Americans operate. But to do this also converts the study of society, from an empirical set of data into a moral test of courage.

With the emergence of an environmental-protection movement, one with great legal clout, all sorts of changes that were in the past taken for granted are now being contested. Typical is a dispute at the University of Arizona in Tucson over plans for a mountaintop observatory. The dispute is between astronomers, who search the stars for the secrets of life, and biologists, who claim that clues to life are to be found within the chromosomes of a tiny red squirrel living in this territory. Environmentalists claim that the construction on top of

Mount Graham would cause the extinction of this dwindling sub-species of squirrel that has evolved in isolation atop these same mountains for eleven thousand years.[14]

Here we have a struggle between two groups of scientists—each of whom are staking claims for the special nature of their subject matter. So often, environmentalism is a class phenomenon, pitting workers against naturalists. But here we have two powerful groups of scientists, each calling up the support of their political leadership. The astronomers call upon the Max Planck Institute and the Smithsonian Institution, while the environmentalists appeal to en-dangered-species legislation and to the Fish and Wildlife Service for protection and relief. Again, without referencing the specific merits of each side, the character of the sides indicates the rooted nature of paradox—neither has any evil intent. At issue is not whether obser-vatories should be moved or squirrels should move naturally, but the very fabric of nature.

When the fabric of nature comes into struggle against the needs of society, we have yet another variety of combustion. It appears that requiring grain alcohol in gasoline might indeed increase the amount of fuel available, and even result in a minor reduction of carbon monoxide in the atmosphere; but it also would create a large increase in smog. Gasahol or gasoline with 10 percent grain alcohol would cut carbon monoxide by 25 percent, but nitrogen oxide, which causes smog and acid rain, would rise by 8 to 15 percent. This works out to 6 percent more smog in the atmosphere.[15]

Such paradoxes have profound impact on the economics of American society. Ethanol is produced by agricultural processors; fuels are produced by gasoline and chemical companies. The intro-duction of gasahol also involves changes in the structure of engines, thus raising the specter of increasing automotive costs to already resistant consumers. Again, the Environmental Protection Agency supports gasahol, but without changes in the automotive industry, this means paying more for dirtier air.

This is not to suggest that such conflicts do not admit of a posi-tive resolution. It is to indicate how complex resolutions of big is-sues can be, even in microcosmic forms. It is, however, clear that not a single proposal for reducing the amount of fuels consumed in auto-mobiles has been put forth that does not entail these sorts of in-creases in monetary costs and environmental hazards. But again, it deserves to be noted that what makes solutions to social problems so difficult is precisely the sense of righteousness brought to the bar-gaining table by both or sometimes many sides. Paradox is a sensitizing agent, but policy is still needed to resolve matters that aggravate and aggrieve.

So many social paradoxes are resolved within legal and judicial contexts because the traditional mechanisms of society have broken down or are atrophied. Norms, habits, conventions, or morals that would have settled matters in the past no longer obtain—certainly not with the compelling force that would permit conflict resolution. Hence, the efforts to move from paradox to resolution of social problems generally take a legal rather than a normative form. However, such resolutions are far less binding on the human actors in the social drama. As we move to larger issues involving entire nations or groups rather than specific epiphenomena, this becomes evident.

A good illustration is the emergence of Indian land rights within advanced nations, such as the United States and Canada. Once land rights are granted, does sovereignty follow? Take the case of the Mohawk Indians, who straddle the borders of New York and Quebec and Ontario provinces. The tribe insists that land rights involve tribal law. That translates into tax-free sales of gasoline, cigarettes, and other items to non-Indians and, no less, to setting up gambling casinos in violation of state and federal codes.

Efforts to restore the Mohawk Indian nation to some condition of autonomy has resulted in massive violence and a breakdown of law as such. The Warrior Society is divided on a course of action, and efforts by officials in New York and Ontario resulted in further violence, with the officials being termed foreign invaders and occupiers. Meanwhile, the Indians themselves are sharply divided as to the value of gambling casinos and the disbursement of the proceeds. There are not only rival Indian factions, but rival Indian paramilitary and police-enforcement units.[16]

At what point does the "law" of the conquering nation (in this case the United States, and more specifically, New York State) supersede the "law" of the newly enshrined Mohawk nation? Are the Indians calling for state intervention really entitled to protection of American or Canadian laws or, having a new sense of sovereignty, stuck with the need to work out their own codification of law and behavior? Thus, what began as a common effort to recognize past injustices becomes in relatively short order a divisive situation involving profound struggles between warring Indian factions as to their obligations to state and federal rulings of the "conquering" nations of the United States and Canada.

Even when there is a seeming consensus that the toppling of a totalitarian regime is a positive good, one notes negative by-products that become ingrained in the new order of things. It has long been understood by experts in developmental economics and sociology that high rates of development may entail high rates of deviance as well. One sees an example of this in the collapse of the Soviet police state. Crime against foreigners in the Soviet Union, now Russia, has dramatically increased as outsiders, long isolated and insulated from communist society, find that their new ability to move freely has exposed them to greater danger.

With the collapse of Soviet power and the corresponding demise of a command economy, crimes ranging from murder to rape to burglary and attempted kidnapping have greatly increased. But theft remains the paramount crime, especially against foreigners. Crimes against the native population have also increased dramatically. Indeed, the tenfold increase of crimes against foreigners is matched by a similar percentage increase of crime in general. But since foreigners have the money and goods that the local population wants, they become special targets. The Deputy Director of the Defense Service Agency, Oleg Fomin, summarized the matter succinctly: "We used to think Moscow was the safest city in the world, but now we know otherwise. And unfortunately for foreigners, they are the people here with the money and the goods that criminals want. Why rob a Muscovite for a few worthless rubles if you can get a foreigner and get some dollars or Deutsch marks?"[17]

The problem is that at a time when Russia, and Moscow in particular, is soliciting Western aid and tourism, such high levels of crime result in U.S. State Department warnings to American business people and tourists or, for that matter, government officials wanting to go to Moscow. Surely, one would not argue a parity of paradox. The seventy-four-year-old totalitarian regime is certainly not going to be brought back in an effort to safeguard foreigners from theft. Still, the fact is that even such an unmitigated "good" as the end to a barbaric regime turns out to be quite mitigated in both circumstances and consequences. And for that very reason, the place of analysis cannot be subverted by mindless celebration.

As a final example of paradox, perhaps one can turn to the case of Manuel Antonio Noriega. The former strongman of Panama was forcibly removed from office by an invading task force of American troops. A certain reduction in the flow of drugs from Panamanian sources may well have taken place in consequence of this. Certainly, the situation in Panama has eased in terms of control over people by the armed forces and the drug cartels. But the questions asked are more about the nature of the trials than the character of the regime headed by the former Panamanian dictator.

Problems piled up from the time of the seizure of the general and his extradition to the United States. Can one put on trial the sovereign ruler of another nation? Can First Amendment guarantees of press freedoms be maintained while asserting the defendant's Sixth Amendment right to a fair trial? As one commentator, Burt Neuborne from the New York University School of Law, put the matter: "Bad cases make bad law. The Noriega case was a bad case from the day they had an invasion to seize him and then seized his assets. Now he has justified the illegal taping of conversations between a lawyer and client, and a prior restraint. Noriega is going to do more damage to the U.S. Constitution than he ever did as a dictator of Panama."[18]

Even allowing this as an exaggeration, the issues surrounding the invasion of a country, the capture and transfer of its leader, and the use of confidential tapes in news broadcasts all raise issues of a paradoxical nature. For even if one assumes that the end of a dictatorship and drug-smuggling ring run with the authority of a state machinery is a pure blessing, opening up the sorts of legal issues that are raised by such actions cannot be dismissed as side shows. Indeed, as the Noriega trial unfolded, the side show became the main show.

My purpose has been clear enough: to show how, in a variety of contexts, drawn almost at random, from a few months in time, the consequence of policies and events, intended or accidental, planned or spontaneous, requires social-scientific appraisal because moral boosterism or ideological flag waving is simply gratuitous, even grotesque. We cannot dispense with social science as long as ambiguity is part of the structure of outcome; nor must we permit the social sciences to become part of an ambiguous legacy by exaggerating their claims and confusing the public as to the limits of their methods.

Social science cannot become a caricature of itself, cannot be reduced to a variety of social physics or abstracted mathematical formulas, since prediction is one and only one small part of the social-scientific enterprise. Predicting the likelihood of a single event may be feasible; indeed, that is what planning is about. But predicting the long-range consequences of any singular event is not simply harder to do, it is often absurd as a goal. In no measure has my purpose been to deny structure to social events, or that knowledge acquisition is inherently limited. Rather, it is to assert that the very natures of both structure and knowledge are laden with ethical judgment. Hence, ethics, and its embodiment in law, are part and parcel of evaluation in the social application of social science.

The social sciences live in a world of paradox because they are a part of, no less than a judgment on, such a world. Social science thus has as a task the anticipation of empirical paradoxes no less than structural consequences. It needs to look upon social conflict as a potential mine field and provide some possible optional courses of action. Its great merit is as a sensitizing agency for behavior, not as a blueprint for action—or worse, slogans about action. The social sciences are a guide to the perplexed. When they become more or claim more (such as a sure-fire method for preventing crisis, bringing about the good society, or re-creating the human form), they become absurd to start with and dangerous to end with. To know the potentials of reason we must accept as a given the paradoxes of experience. To predict the range of consequences to a particular action may be less enthralling than simple linear regression analysis might permit. But it might move us to a higher scientific ground—one with deep human potentials. For it is the very fact that social knowledge is bathed in paradox that permits human beings to constantly move back and forth between empirical analysis and moral reflection. In short, paradox is at the heart of the human soul; and the study of human souls is what social science is all about.

This is why social science must keep a distance from political battles of the moment. Max Weber knew all too well that "the final result of political action often, no, even regularly, stands in completely inadequate and often even paradoxical relation to its original meaning."[19] He also knew that politicians must always have faith in the causes they advance. But truth, as social science pursues it, and faith, as politicians have in their causes, are like oil and water—they do not mix. In fact, they stand in paradoxical relation to each other. What Weber called the "paradox of unintended consequences" is what politicians suffer but what social scientists, at their best, study.

13

Freedom, Planning, and the Moral Order

American notions of freedom, planning, and democratic society are largely derived from a twentieth-century context and, more exactly, from a post–World War II series of notions that emphasized a belief that a caring society must overcome a careless state. The operative word in this, of course, is planning, since we obviously derive our notions of freedom and ethics from a large variety of sources ranging from ancient Greece to eighteenth-century Europe.[1]

In the United States, the presumption of equity was set in the 1930s by the New Deal at essentially subnational or community levels. Its aims, in retrospect, were less to destroy a wealthy social class than to enlarge the stake and participation of the working classes. Taxation rather than expropriation became the essential mechanism for reorganizing society without destroying the established social order in the process. In this, the fifty-year period from 1933 to 1983—when the Reagan Revolution took full command and went into high gear—represented a remarkable ideological continuity; a consensus about social goals unlikely to be repeated easily, given present schisms within the economic fabric of American life.

Planning took root in every American village and hamlet. Indeed, given its basis in issues of housing and zoning, planning was more common a feature in rural and suburban areas than in the older, industrial zones. As I have noted in an earlier essay, this was diametrically different from Soviet notions of top-down planning, in which it was precisely industrial location and productivity that was to be measured. In the United States, planning was linked to community. In the Soviet Union planning was a function of the state. The differences in inputs and outputs are now part of the tragic history of the century.[2]

As a practical matter, then, bottom-up social planning, in contrast to top-down economic planning, derived its force from the catastrophic series of wars that left Europe in ruins and permanently shifted the focus of world power—not from capitalism to communism, but from Europe to America. That the United States saw fit to rebuild the defeated powers of Germany, Japan, and, to a lesser degree, Italy was a function of economic farsightedness and not just of ethical concerns. Capitalism began life as a world system and continues to derive its strength and values from remaining a world system. If it takes a greater amount of noneconomic intervention in the affairs of the economy, that becomes the price to be paid for system maintenance as such.[3]

The second half of the twentieth century witnessed growth not only in the amount of real wealth but also in the attention focused on the disparities of wealth and poverty. It did so on a global scale that in the past was readily ignored. But such *laissez-faire* attitudes could no longer be maintained given the nature of advanced technology in communication and transportation. The major powers agreed that the price of the defeat of fascism would be not just a restitution of destroyed classical economies but also a redistribution of wealth within those systems. This took place among the victorious Allied powers no less than among the defeated Axis powers.

This strategy clearly had successful consequences. In retrospect, one must note how remarkably stable the locus of power has been throughout the century: while relative weights shifted, the United States, western Europe, Russia, and Japan have remained central to world economic power. Those new players that have emerged (such as Canada, Brazil, South Africa, and the Pacific Rim nations) did so largely in the context of capitalism and free-market systems; but they remain as nations with relatively high social-welfare and local-planning components.

It must therefore be appreciated that when I speak of the drive toward equity, or even of administrative mechanisms like planning to achieve economic parities, I am talking of developments within the capitalist context. The warfare between capitalism and communism essentially did not take place because of a single overwhelming fact: the near-total breakdown of communism as a dictatorial system that could not deliver the social goods it promised. If I belabor the dismal performance of alternative systems like fascism and communism, it is to make a point: that examinations of equity, planning, and moral purpose no longer occur between systems, as we were all led to believe would happen, but within the bowels of capitalism—a system repeatedly consigned to the ashes of history by Marxism–Leninism. With this rather startling shift in the locus of social relations understood, I can move forward to an examination of the routinization of democratic theory and its moral consequences.

The drive toward universal equity had multiple sources that predated World War II. In fact, this impulse became powerful after World War I. Social-welfare advocates, such as Keynes[4] in economics, Mannheim[5] in sociology, and Laski[6] in political theory argued that individualism need not be sacrificed in the drive toward an equitable society. Keynes gave expression to this tightrope between personal liberty and collective responsibility when he wrote that "the important thing for government is not to do things which individuals are doing already, and to do them a little better or a little worse; but to do those things which at present are not done at all."[7] For these egalitarian theorists, the good society implied a caring concept; which in turn entailed a distributive concept of social welfare. The management of goods and services became the essential moral ground of the benevolent state. In this way, planning was absorbed into capitalism, without adopting revolutionary options or class antagonism as a political goal.[8]

While recognizing this democratic impulse of socialism and welfarism in the West, it nonetheless shared with the totalitarian regimes of Communist Russia, Nazi Germany, and Fascist Italy a belief that government must be central in the redistribution of wealth and resources to the overall needs of the society, and that these needs are determined by the state. The British planning model became an archetypical example of how to resolve problems of stratification on one side and economic fluctuations on the other. The Swedish approach went further, in that planning reached into the inner workings of family life as well as into economic planning in general. In the United States, urban planning served the same larger social goals, yet retained a much higher level of individualism than either the British or Scandinavian approaches.[9] In the Soviet Union, it was the presumed leveling potential of the industrial system that would have egalitarian consequences. To this end, the entire middle class and rural peasantry was sacrificed by Stalinism. In Nazi Germany, a combination of state and military planning provided the catalyst for mass mobilization and heavy collectivization.

Liberal and radical regimes alike appealed to neo-populist leveling sentiments. They also made the moral presumption that the selfishness of the older ruling classes could only be curbed or crushed by external forces: the military, the police, or the state bureaucracy. Curiously, the idea that the so-called popular classes could also be selfish and venal did not occur to many of the radical theorists; nor did these same figures stop and pause in their commitment to a state determining the full range of human emotions. The most democratic economists, such as John Maynard Keynes or Gunnar Myrdal, distinguished between the things individuals do best and the things that collectivities do worst. But neither of these figures informed us who would make this rather fundamental determination in everyday political life. As a result, far from resolving classical antinomies between liberalism and conservatism, they were simply left smoldering under an avalanche of unmet social needs; surfacing at a time and in places where (and when) such fundamental needs were met.

Little wonder, then, that with such a statist consensus, a barrage of planning instruments become dominant for most of the century. Conservative doctrines of free markets and *laissez-faire* orientations fared poorly and were constantly on the defensive. William Mallock[10] and Walter Lippmann[11] and later Milton Friedman[12] and Friedrich Hayek[13] seemed churlish if not plain backward and reactionary to argue the case and cause of individual liberty in the midst of a depression prior to the war and decimation after the war. The electorate of America and Europe concurred. Thus, planning became a fixed star in postwar American and European reconstruction and beyond. The only cracks came with a set of problems that arose after planning modes became enshrined as *de rigueur* in all advanced nation-states.

Those cracks, which by the present generation have become yawning caverns, are readily summarized: first, the loss of innovation in advanced societies by scientific and technical personnel; second, the erosion of personal savings and expropriation of personal property; third, the corruption and partial collapse of banking institutions that counted heavily on state and federal supports; fourth, the emergence of an underground economy, or the sheer expansion of political corruption as a method to beat the high costs of a planned economy; fifth, the rise of heavy taxation to pay for planning programs that were deemed to be socially necessary and that led to a struggle between working classes and welfare classes; and sixth, the creation of bureaucratic and administrative sectors that, far from doing away with class antagonisms and stratification, only served to change the character of competition for the control and disbursement of scarce resources.

The climate of opinion changed. It became a canon that the democratic purposes of planning, far from limiting the centrality of the state as an oppressive mechanism, only served to increase bureaucratic management and enlarge its physical size.[14] In the process, liberal and radical critics of state power of the nineteenth century became defenders of the state in the name of saving the plan. The conservative critics curiously became the torchbearers for a reduced role of state power, especially in the welfare arena, arguing the premise of community rather than order. It was not the new conservatism, but the new liberalism, with an active view of state power, that changed the image of politics from an entity based on a struggle for representation to one based on control of the levers of economic power. In short, as the century wore down to its conclusion, it became clear that the issue was the struggle not so much between *laissez-faire* economics and planning mechanisms as between community and centrality; that is to say, the extent to which communities, however defined, determined their own destinies and to which central administrative regimes, whether in communist or capitalist zones, determined what those destinies would become. All sorts of fault lines shifted. Liberalism contained the seeds of its own destruction. It shifted from a doctrine of that government governs best that governs least (that is, a Lockean mode) to a doctrine that that government governs best that determines the appropriate range of wealth and poverty (that is, a Laskian mode).[15]

The main purpose of my remarks, however, is not to restate the abstract arguments between moral philosophers and political pundits on the limits of authority and the needs of justice but to reconsider anew, in the light of current circumstances, just what are the efficacies of planning mechinisms for the decade ahead, that decade which will take us to a new millennium. We need to get beyond panaceas and panoramic visions and parcel out what we mean by justice in a context of democratic planning. In this context, we need to know that the classical Greek concept of justice had to do primarily with the individual, that is, with the just man; the modern Anglo-American concept of justice has to do with the just society; and, added to these, the Continental notion of distributive justice has to do with the presumed obligation of the government to the citizen, or the just state. In this set of relationships between individuals, societies, and states, we have to come to terms with moral options in dealing with everyone from the opulent to the homeless.

It is widely recognized that justice is not a fixed moral category but a changing political goal toward which people and institutions apply differing value judgments. One can be a just person in an unjust society, or as Reinhold Niebuhr put it, "a moral man in an immoral society." But one can also be a criminal in a presumably just state. One can be alienated from a tightly knit community or indifferent to a just society. Indeed, one of the enduring lures of the liberal state is its high toleration for deviant, that is, unsponsored, behavior. Even to admit such options opens up new possibilities for discourse on the moral meaning of planning in a new environment and in changing circumstances. But it should also serve to caution, perhaps undermine, faith in state authority as such.

The notion of planning is now subject to intense scrutiny. Urban planning in the United States, while indeed not the same as the industrial planning of the Soviet Union, still makes demands for change. For starters, the United States features planning as a community concept, a regional concept, and, above all, a partial concept. The Soviet Union, at least until its present disintegration as a communist state, featured a concept of national planning, based primarily on industrial productivity rather than personal proclivities set by state goals rather than social mechanisms. That, at any rate, was the case prior to the collapse of communism and the emergence of a modest form of pluralism in the U.S.S.R. in which market mechanisms to regulate the economy emerge.

Notions of planning can be quite diverse, being applied at times to the family economy and at other times to global systems. This was done for the purpose of seeing to it that the members of a nation should survive intact. But whether past notions of planning were partial or holistic, they had in common an appeal to serving the needs of people through political mechanisms. At the least, the planner is one who keeps the clock striking the right time; but taken to an extreme, the planner defines the nature of time itself. As a result, planning became central to ruling statist notions wherein systems of policy science run by experts displaced elected officials as sources of legitimation.

Underneath much discussion of planning is not only the question of whether a society can or cannot be just outside the plan, but also the issue of at what point inequalities of housing, education, or health are tolerable. For as Rawls put it in his influential *A Theory of*

Justice, the moral worth of a plan is not only economic fulfillment but also social equality.[16] The force of planning agencies at the macro-level has, in the past, been their ability to mobilize and organize the collectivity to reach certain ends held desirable by hierarchies and elites acting in the name of the people. Far from being an open and shut matter, this must be done through strategies and tactics for the preservation or destruction of a bourgeoisie or a proletariat. The issues raised by planning go to the heart of present-day institutions and ideologies in both the West and the East. In this sense, planning is far more than a device, it is the common denominator on the field of social action of twentieth-century social theory. Planning organizes, in social practice, the moral presumptions of economic entitlement.

The counterrevolution against excessive central planning quickly spilled over into a direct attack on the moral foundations of planning as such. And in the work of Nozick, the sickness of planning places a cap on such vulgar economic notions as distributive justice.[17] For what began as a social assault against the absence of a caring system, ended as a statist attack on the private person. Equity was transformed into leveling, and liberty became a concept used by conservatives to stem the tide of social welfare by upholding the role of the individual.

The effort to move away from the abyss of total regulation became the touchstone of the revolt against totalitarianism in the communist regimes and, even more profoundly, the revolt against taxation in the capitalist regimes. Everything from a free-market sector to stock-market organisms are introduced to check the abuses and excesses of the planning mode—even as the machinery for political repression is increased. In the United States there is growing realization that urban planning without commercial zoning is senseless in an environment in which industrial parks are places in which people live no less than work. By the same token, Russia has taken into consideration supply-and-demand ratios and direct measures of public opinion on everything from commodity quality to alcohol abuse in order to determine priorities. While the "free market" remains largely unexplored in the newly evolving regimes of eastern Europe, "free ideas" have become the touchstone of Europe as a whole.

The entire post–World War II environment to which we have grown accustomed is now under intense scrutiny. This review is global, since it is apparent that goals of perfect equity are not easier to achieve than they were forty or fifty years ago. While at the earlier period (the halfway mark of the century) the benefits of planning were being uncritically extolled, now the costs of planning, the social and political costs no less than the economic costs, are being critically examined. In such areas as zoning for new housing, busing children from one district to another, making the streets available to the dispossessed or safe to the possessed, the issue of the plan or the absence of such is simply a device to permit moral choice to go forward in a legal context.

We should begin with a simple proposition: plans work when people work. If planning becomes an overbearing or overarching model imposed from above, people will not work, or will work only if it is "off the books" and, hence, beyond the planning context— part-time and uncreatively, just enough to get by on. Totalitarian schemes of planning, thus, invariably involve coercion and threat no less than populist appeals to equity and fairness. People have not become mean or nasty in any Hobbesian sense, but as the planning notion has become institutionalized, the pressure to "get results" has increased. It is now realized that the rationalization of wealth, goods, and services carries high costs. Hence, a former end in view, the plan is transformed into an instrumentality. It is now subject to a means test; that is, to the pragmatic regard for the limits to policies and principles.

The social-indicators approach, for all of its crudities as a measuring instrument, served to remind the professional policymakers that there is a multiplicity of bottom lines. Questions of a new sort began to be raised by those in allied social sciences. Has planning led to a breakdown in personal courtesies to one another? Has planning led to market confusion between key segments of society struggling to secure advantages or personal gain? Has planning produced a set of evaluations to permit nonpartisans to measure the gains versus the damage of particular plans implemented?

Increasingly, the problem of planning is structured not so much as planning for whom but planning for what. Is the goal of a plan itself growth or maintenance of present privileges? Can one plan for no growth with the same intensity as for growth? Again, the automatic set of assumptions behind planning theory has broken down beyond recognition. In the West, the idea of competition has had a renaissance. Increasingly, in the Chinese and former Soviet orbits the idea of competition between planning units has become acceptable, even preferable to monolithic, industry-wide planning agencies.

Certain features of planning, those connected with exaggerated claims and prerogatives of the state, have clearly been pushed back, if not entirely discredited and eliminated. Among the victims of this new-found sense of the paradox of planning are the following myths: planning as a mechanism for the abolition of social classes and inherited distinctions; planning as a mechnism for punishment of enemies of the state; planning as an efficient long-range forecasting device (that is, five-year plans, ten-year plans, etc.); planning as a substitute for budgeting and financing within the parameters of what a nation or community does best; and planning as a political mobilization tool, a device to create public support for nasty or faltering regimes.

There is a growing recognition that wide segments of social life do not require and indeed suffer from excessive top-down regulatory agencies. Such agencies add a layer of bureaucracy that increases costs and decreases innovation. Even if one rejects concepts of self-regulating marketplaces or utilitarian doctrines of the private will adding up to the service of the general will, it remains the case that personal spontaneity and freedom of action, without punishment for failure, are themselves aspects of a just society, no less than is adjudicating different claims a characteristic of a just state.

The appeal to planning has, in the past, often been a defense of egalitarian models of social arrangements. These have become less appealing as it became plain that state planning commissions and even local and county and state planning bodies themselves stimulate inegalitarian instrumentalities and are often tyrannical in their execution of the plans. The struggle around such legalities, mandated items such as school busing, housing start-ups for middle-income groups, and even salary caps for athletes, come upon individual wants and needs to make of private property a personal fortress against social plans.

The notion of planning as a mechanism for bringing into existence a new man or a healthy community is also less likely to survive the present era than previously imagined. The fallacy of assuming that economic change or technological developments move in lock step with moral probity has been finally and fully punctured in all but the most backward societies. The fact that we still employ Platonic and Aristotelian notions of the good, the right, the beautiful, and the just, long after we have discarded Plato's doctrine of mathematical numbers or Aristotle's doctrine of the spontaneous generation of species, indicates clearly enough that different realms of reality exist, moving at different rates and servicing different ends.

Planning will not dissolve as easily as the Soviet Empire has collapsed. But it does become more modest, more tailored to actual needs of a population. Democracy is not proven by the extent to which a society has planning commissions; rather, planning commissions are proven to be democratic by the extent to which they are responsive to the pushes and pulls that summarize the common good. Planning thus becomes increasingly short-ranged in order to remain viable. The number of factors, the number of variables taken into policy accounts, have exploded. As the universe becomes interdependent and the world smaller, the potentials for long-range plans diminish. Artificially or politically induced petroleum and energy crises totally destroy prospects for planning in the abstract. Overseas competition in a given industry reduces the ability of a nation to plan its long-range budget. Estimates of General Motors' growth are curtailed not so much by changes in consumption of finished goods, in this case autos, as by the overseas competitive sources of such finished goods.

Microplanning replaces macroplanning as the touchstone of the 1990s. Fine tuning differences displace demands for total overhaul of ecologogical systems. The danger in this is that the period may be so short, months rather than years and years rather than decades, that crises, or at least radical swings in economic stability, become a permanent part of the economy and polity of highly advanced industrial and urban nations. In other words, the cycle of crisis and crisis management is repeated rather than resolved, and the demand for the planning body is re-created, precisely because of this weakness in forecasting and budgeting long-range trends with accuracy.

Planning enters into the cycle of social life as a process of guiding change and as a system for regulating change. It is neither a panacea to guarantee equities in housing or health nor a "new moral law of consciousness" governing society as a whole as Soviet socialists once believed. But neither is planning a Draconian imposition on the modern world, a new form of modern slavery. Such overheated rhetoric disguises the need for post-industrial societies to maintain their services in opposition no less than in support of the state. Creative tensions do not dissolve; nor are they overwhelmed by exaggerated responses by planning theory.

Increasingly, planning is part of the common heritage—an activity performed by policy experts at a variety of levels (industrial and urban, and by big and small industries no less than big and small governments). There is a recognition that the failure of a specific plan is not the same as the failure of a society. Indeed, were they isomorphic, the Soviet economy, given its catastrophic series of five-year plans, would long ago have vanished into outer space. It takes many economic blunders to finally topple a polity.

The success of a planning body is not the same as the achievement of a just society. Indeed, implementing plans involves institutionalizing new bureaucracies. Increasingly, it is recognized that justice is a multifaceted concept and that a certain amount of economic expansion and income disparity may be a prerequisite to psychological drives to work hard and participate in the common good through service to self and family. Good housing or excellent health care, whatever that implies, are not rights of birth, not entitlements. Quality housing or care is a shifting sand—but it is made so by ever higher standards of the good. Neither egoism nor altruism are fixed moral poles; hence, neither the planning body nor free human space can be judged as fixed social goals.

New permutations and combinations in Third World societies— Leninist single-party states which advocate free-market economics; democratic varieties of socialism in which punishment is replaced by positive reinforcement; multiparty states with centralized economic units—are taking place in an experimental framework which recognizes that the end of old ideologies is at hand. With the disappearance of these ideologies, one can witness the return of moral discourse based on classical notions of individual and social goods, personal and general wills, and private wants and public needs.

This is another way of saying that changes in the moral structure have greater utility when stripped clean of ideological ballast and timeless aims. Planning is more efficient when divested of the frightening authoritarian models that formerly enshrined and ensconced it. This is not to insinuate a free-market vision of a future based on purely voluntary consensus, a return to vague utilitarian models of the general good as a function of particular goods. It is to assert that the age of total social planning is clearly finished. No longer do large numbers of people, or at least wide sectors of leaders, think that planning can correct vast social and historical injustices and income imbalances. Too many people have been punished by too much fanaticism. Instead, planning is viewed as part of the storehouse of policy tools. If there is less talk about the good society, there is, I trust, more appreciation of micro-management as one of many devices to reach out for a better society. This, at any rate, is a personal view, perhaps hope, of what we can expect of planning theory and practice for the remainder of this century.

Over the course of the twentieth century, the notion of planning itself has passed first from the realm of ideological discourse, then into that of political dogma, and finally into that of social-scientific research. If we compare the current period with the situation ten or, at most, twenty years ago, this glacial shift in attitudes and orientations will become more readily apparent. In the United States, after more than a decade of Republican-party administration, the talk about market mechanisms that are self-regulating and self-correcting has substantially abated. Earlier notions of morally centered equilibria have yielded to a far more moderate mode—shared by both major parties—in which it is recognized that the parameters of such issues as health, education, welfare, and housing will not be automatically self-adjusting for the total American population.

Supplies and demands may seek, but rarely discover, equilibrium points. Quite the contrary, current practical wisdom in America is that while a majority of the people, safely ensconced in the economic process, can employ their earnings and profits to fine tune their own needs, the same cannot be said for a present and persistent underclass with no such potential. Further, the common wisdom is that some sort of social planning is required to diminish if not entirely dissolve this underclass. Further, devices such as taxation, income redistribution, and resource allocation are now seen as residing within a limited planning mode.

Within the former Soviet orbit, dramatic, even astonishing, changes are under way in communist practice. The inherited idea of psychological predilection among people that planning is a cure-all for everything which ails an economic system has simply fallen by the dialectical wayside. The inability of the Soviet economic order even to catch up much less surpass the economies of the West has led to the demise of the Soviet bloc itself. Russian and eastern European constitutional reforms included large free-market zones, greater emphasis on local participation, autonomy in decision making within the economy, legal safeguards for personal safety and private property, economic doctrine which takes into consideration everything from equipment obsolescence to the costs of innovation, and even the advantages of unemployment to adjust market needs. To the relief of the Russian people, older Stalinist notions of population mobilization, consensus-building through the intensification of class struggles, and top-down management pursuits have given way to more modest goals with more realistic means of achieving them. This very shift from the supposed moral management of a society instilled by an ideology permits genuine ethical discourse to flower.

Planning will not dissolve in the Russian or eastern European context; but it will be determined by people needs rather than presumed geopolitical requirements.

Planning for the new century has already passed from the heated demands of politics into quieter channels of administrative maneuver enshrined as public choice theory.[18] Even if such new doctrines are not theoretical panaceas, they point in heavily democratic, bottom-up directions.

This is not to suggest that planning has become a toothless tiger or without moral meaning or merit. Quite the contrary. There is a renewed sense on the part of major powers and systems that the purposes of planning are not the enlargement of planning boards but increased human welfare. The terms of debate shift rather than abate. Strategic considerations come to prevail over systemic demands. Which social system can best utilize its resources to produce widely satisfactory outcomes becomes critical and itself a matter of open moral discourse. Types of planning and limits of administrative control replace earlier models of categorical all-or-nothing approaches. There will still be plenty of issues to debate and goods to demand. But as system survival becomes once again central, strategies of planning rather than principles of economics become the touchstone for selecting policies and politics for the balance of the century.

Essentially, what I have herein charted is the evolution of the means–ends continuum in the context of ever increasing social responsibilities for the indigent or poor. For most of the twentieth century, populist and radical doctrine dictated that planning for economic redistribution was the highest expression of "justice." This has now given way for all concerned. In the East as well as in the West, planning in housing, in health, or in industry is basically an instrument, a mechanism to achieve desirable human outcomes. Models of progress and, worse, sheer devices for "catching up" to advanced nations are not ends unto themselves. Techniques of domination are improper instruments of equity. When planning becomes an end-unto-itself instead of servicing an end-in-view, then organization of a society becomes utterly confused with individual goals and thus serves to confound the nature of instrumentalities as well. Much blood, real as well as symbolic, has been spilled to arrive at a modest, authentically pragmatic sense of human purpose. If a more modest and less certain sense of techniques of rule were to have guided the century, liberalism would not be in its current slump and conservatism would long ago have found a useful role inside rather than outside the social system.

It has been the unique role of social science to remind the policy planner that people need protection not only from the design of discord, represented by the open marketplace, but also, no less, from the added layer of administrative bureaucracy required by the planners in the name of collective order and social solidarity. In the attempt to steer a course between the anarch and the behemoth is where social science and public policy meet. The emergence of public choice in decision making, especially in the area of the economics of allocation, has solidified this new set of relationships. Such a fusion of social science and public policy may well represent the high-water mark of *fin de vingtième siècle* reasoning about human goals in democratic societies. Whatever the authentic risks in entrusting vital decisions to a class of appointed (sometimes self-anointed) as opposed to elected officials, this, at any rate, holds out the potential for a more powerful and beneficial sociopolitical system than the dictatorial imperatives of the *fin de dix-neuvième siècle* revolt against reason. East, West, North, and South all seem finally and fully to have transcended the fanaticisms of ideology. It is the historic duty of, and opportunity for, the social sciences to recapture at the end of the century the open-ended world of the pragmatism of William James, John Dewey, and George H. Mead with which we began it. Or put another way, social science must take into account the fact that fanatics and ideologues, following what Weber called an "ethic of ultimate ends," have given way to cooler-headed politicians and planners who are most likely to follow an ethic of proximate responsibilities.

14

Social Disputations and Moral Implications

I have great qualms about the notion of moral education serving as the ground upon which social research walks.[1] At best, it is a nineteenth-century concept smuggled into sociological discourse by those who search for the City of God in the cities of men, and, at worst, it is a late twentieth-century concept appealing to fanaticisms of the worst sort. Clericalists tend to talk of moral education; secularists, of education for moral choice. The kind of unease or discomfort with the concept of moral education herein expressed is not simply personal but one felt in general among serious people.

The great thrust for reintroducing moral education derives from sources that are usually alien to the intellectual environment of the open university or the democratic persuasion at large. Reflecting upon fundamentalisms of all sorts—Christian, communist, and conservative—the demand for moral education seems to be a mandate which has little to do with democratic education and much to do with dogmatic instruction. In such a society, everything from truth to discipline "involves the idea of obedience to authority."[2] There is a long thin line of scholars, from Auguste Comte to Émile Durkheim, who shared in such demands of linking sociology to morality. That it has not been entirely erased is reflected in the very need for such a chapter as this.

Many scholars in the academic world, especially in naturalistic philosophy and the social sciences, are appropriately ill at ease with the notion of moral education. The phrase implies (or better, disguises) much more than it can deliver. It implies a foreknowledge of what constitutes the moral ground or the normative base from which an understanding of right and wrong derives. From Plato to Durkheim normativists have written with a definite sense of the ideal social and political orders and of the place of moral behavior therein. For the most part, social philosophers did not come to moral theory in terms of a doctrine of choice or a notion of voluntary decision. Quite the contrary. They came to social questions with powerful and oftentimes fixed apriorities. In Deweyan terms, theirs was a quest for certainty rather than experience. In diametrical opposition to the pragmatic rebellion against the nineteenth-century absolutes of organicism in social science and idealism in philosophy, those who have renewed the quest for moral education sense what it should be. Moral behavior is not experienced but is an entity mandated by the heavens, by history, by biology, or, as is more often the case, by metaphysics, by metahistory, or by metabiology.

In the twentieth century, European social theorists like Jacob Talmon,[3] Hannah Arendt,[4] Elie Halevy,[5] and George Lichtheim,[6] among others, have understood much better than their philosophical or psychological counterparts the problem with an explicit code of ethics or a moral education; namely, it carries within itself, no matter how genteel or clever, the totalitarian temptation—a notion of knowledge as foreknowledge. Those for whom a democratic environment remains a pledge no less than a style must pause and think carefully about the vision of moral education; above all, its instructional mandates emanating from providence, history, or biology. The best way to do so is to move from the mellifluous tones of moral education as a general doctrine to the harsher tones of the moral educators and their specific message.

The prolegomena announced, what next? Suppose there is an appreciation of the risks entailed by the claims of moral educators; where, then, do we go from here? It is certainly insufficient and inadequate to make a pronunciamento as if it were an analysis. The issue remains, how is the democratic persuasion transformed into moral doctrine? More pointedly, can such a transformation be either observed in the history of ideas or urged upon a present generation in search of a firmer footing than the shifting sands of pragmatism but which prefers its ideas not to be cast in stone?

One course to take is represented by a theory of pure choice, which pushes ethical theory into empirical examination. This would transform ethics into a question of decision making or value theory and ultimately into a kind of laboratory of mechanical engineering—what ought to be done to bring about desired ends in terms of available strategies, tactics, and techniques. Morality thus becomes reduced to engineering modes and principles; to concepts of how to manage, manipulate, and massage reality at various levels of human interactions. This interactionist approach is uncomfortable for many, since it simply postpones a settlement of moral accounts by drowning ethics in an interactionist pool, and judgment itself is invariably abandoned. What other options are then available? Let me start with a review of how moral education has been achieved in our century.

My presupposition is that the pragmatic rebellion against authority has been so thoroughgoing that education as such could not admit the concept of morality through its front door. Relativism became not so much a choice of values as a long vacation from values. But pragmatism could not avoid allowing moral education to enter ("smuggled" might be a better term) through intellectual back doors. In fact, its emphasis on experience and evidence invited such an approach. The relativistic persuasion was itself converted into a moral canon. Until recently, the word "morals" was considered impolitic; having yielded its princely seat to the word "ethics," with its less instructional and more metaphysical presuppositions.

I am not trying to prove either that there was a pragmatic–liberal hegemony from World War I to the recent past or that the fundamentalist response has only now crystallized. Neither proposition is correct: pragmatic–liberal approaches to social research were dominant, but not unanimously adhered to. Likewise, the fundamentalist–conservative response never quite died out, and from the early 1950s it made steady inroads on social doctrines. What is the case, however, is that the questioning of the liberal consensus has grown into a crescendo of firm intellectual opposition. This quantitative shift has grown to a point where qualitative changes in the nature of social science itself are involved. In this sense, issues of moral judgment and education are directly impacted; since the empirical rationale of ethical precepts are profoundly affected by such paradigmatic shifts in social research. We must now review how social-science disputations have served to underscore current moral standpoints.

Let me briefly review how issues of moral education in several fields of social science are treated. The field of sociology, specifically in terms of social stratification, offers a central starting place. How does it treat the notion of moral education? Essentially by indirection. Moral education is implied in a theory of stratification, which holds that the goal of equality of opportunity and starting points is central to a good society. Sociologists strongly suggest that there is a good; and it is operationally determined by the maximization of equality. Sociologists began to compile data and engage in field researches which demonstrated that vast inequalities abound in American society: racial inequality, religious inequality, sexual abuse, child abuse, and agism. This is, after all, the real message of *Middletown* and *Middletown in Transition*—that is, both the *structure* of inequality in the former and the movement toward equality in the latter.[7] The entire area of the sociology of stratification was dedicated to statistical enunciation of such moral truths. What appears as fact (namely, data on inequality, or differences in earning power and occupational roles between males and females) became readily and rapidly transformed into a statement and judgment about the need to remove inequality.

Sociology strongly implied that within its framework, data on stratification are gathered and verified but that the very enunciation of such data provides bona fide evidence for the removal of inequality and a firm movement toward perfect equality. This is essentially to be brought about through legal means, political techniques, and labor organizations. But the strong and lasting impression of most empirical studies of stratification is that one should work toward a society built on equality and not simply accept passively the description of societies that have built inequities. There may be powerful outstanding differences as to how successful the United States has been in its pursuit of perfection, but there is little serious debate over equity goals as such.[8]

If we move briefly to the area of political science, the same kind of overwhelming judgment about egalitarian goals is observed: all people should count as one, no more than one, but no less than one. Therefore, in the study of political science—whether it be pluralism in the practice of political systems, decision-making distribution on the Supreme Court, the checks and balances of Congress, or even the economic characteristics of voting behavior and mass participation—political scientists have been virtually united in their idea that the politicized individual should count as one, no more and no less. Voting-rights legislation and congressional redistricting schemes both are deeply rooted in feelings along those lines. In the words of E. E. Schattschneider, "democracy is a competitive political system in which competing leaders and organizations define the alternatives of public policy in such a way that the public can participate in the decision-making process."[9] Equality of rights is constitutionally mandated. Recent Supreme Court interpretations have constantly argued the need for maximizing opportunities as a basic mechanism for ensuring rights. Political science presents a continuing effort to pierce the disparity in the employment opportunities of races, the political opportunities of sexes, and the business opportunities of different religious and ethnic groups with the omnipresent ideals of a democratic social order.[10]

A fascinating illustration of how egalitarian modalities became dominant in social science is in the area of criminology. Far from the late nineteenth-century tradition that sought to indicate how the behaviors of criminals differed from those of other citizens, the new criminology took the position that criminals are essentially involved in the redistribution of wealth and privilege. Hence, in its most recent version, crime is seen as a species of economic behavior or even recreational behavior. Crime is viewed as expressing a grievance and, hence, belongs to the same family of social behavior as gossip, ridicule, and punishment. Ultimately, crime comes to be viewed as one more type of behavior that helps to ensure, not undermine, social control.[11] Crime is simply viewed as a normal way to evolve a doctrine of self-help. In such a pure democratic view, it is the continuum between criminal and non-criminal behavior that is central, not the nature of the criminal element. The very distinction between normal and criminal behavior is reduced to the whimsical nature of law in society. Relativism thus ends, not in moral choice, but in sheer intellectual nihilism.

Anthropology, given its powerful relativistic bias, has also contributed mightily to this same egalitarian impulse. Not only should every individual count as one, no more and no less, but also civilizations should count as one, no more and no less. Cultural anthropology in emphasizing differences between societies has also inadvertently argued against claims of moral superiority or moral inferiority.[12] There are differences between cultures; there are distinct patterns of behavior; but these cultural patterns are built up over long periods of time. Non-industrialized societies cannot simply be described as being wrong; nor advanced industrial systems as being right. The nineteenth-century distinction between primitive savages and modern civilized beings is an impossible hypothesis to sustain. Bronisław Malinowski was typical in that he always managed to poke fun at the notion of being civilized, while elevating savagery as a legitimate economic system in its own right.[13] His references to civilized peoples aim to convince us that the empire builders engage in a higher form of savagery unknown to most of his beloved Trobrianders. The way Englishmen behave at soccer matches, in crowd scenes, or with one another belies the notions of civility and civilization. There is little doubt that for most cultural and social anthropologists, empire builders are not only no better than primitive peoples, they are much worse. They carry within themselves the possibility of much greater destruction over a much wider geographic terrain.[14] There is, furthermore, no doubt that the founders of cultural anthropology shared this deep bias on behalf of the less developed and against the most developed. Anthropology carried within itself not only the presumption of equity but also the need to learn from other peoples. As a consequence, the strong relativistic impressions and egalitarian impulses are just as powerfully etched in anthropology as they are in political science and sociology. And such relativism carried the full weight and authority of ethnography, of fieldwork in far-away places.

If we turn to economics, the main practical revolution of the twentieth century, whether it be of a Marxian or Keynesian persuasion, is heavily informed by the same egalitarian direction. Setting aside momentarily the competitive theories of scholars like Friedrich von Hayek and his more recent followers,[15] the main impulse of econometric models has clearly been the idea of a social floor and an economic ceiling.[16] The central presumption behind the post-Keynesians is to get beyond the equilibrium, beyond the world where there is no past and future and no history, and into a world of steady growth. Only such management of growth permits safety nets that do not tear and ladders that do not favor the haves over the have-nots.[17] Equilibrium is thus a function of growth, not stasis. And this means bringing ever increasing numbers into the economic system as equal players.

There is a utopian element in this quest for economic equality, but the wide variations between Marxian and Keyesian concepts notwithstanding, the shank of the quest is unambiguous and shared by both schools of economic science: to create a social order in which personal opportunities on a level playing field are created by state power. Economists for the most part have argued that people have entitlements. These entitlements come not from work but from being human. This is a radically different notion from the Ricardian free-market faith in subsistence or the Smithian notion that the degree of wealth derives only through what a person contributes to society. Quite the contrary, the dominant mode of twentieth-century economics is that a standard of living is an entitlement by virtue of being human. Thus, even the economists, whom one might expect to be the toughest and roughest of customers on questions of cost-benefit analysis, are sentimentalists. They fervently believe that fiscal and monetary policies should be actively and continuously deployed toward achievement of full employment and price stability. Not free markets, but total employment became the ethical cornerstone of modern micro-economics.[18]

The area of psychology that has been most sensitive to this revolution in attitudes toward inequality and discrimination has been abnormal psychology. The strong language of moral disapproval for sexual difference characteristic of European founders of the psychiatric movement has given way to a neutral rhetoric bordering on outright approbation of different groups. What was formerly labeled infantile disorders, or the inability to transfer affections from the same sex to the opposite sex, has now become simply a matter of alternative life-styles or the ability to love or the capacity to create.[19] Optional modes of behavior displaced fixed norms as the key to the analytic quest for normality. Patterns of behavior once deemed deviant are now increasingly perceived as nothing more than an inherited bias of the professional establishments. Authority itself has come to be viewed as the enemy of mental health; to be replaced by a world in which the analyst and patient both become equals in a sharing support group, which in turn will evolve into communities and groups of different types prizing individuality and free expression.[20] Most recently, the trend has been adjusting to such life-style differences as lesbianism and homosexuality rather than urging individuals to find their way back into a vague social mainstream. This is yet one more illustration of the same theme witnessed in each of the social sciences: relativism in behavior rather than absolutism of moral claims; rights to those who choose to be different or cannot help themselves from being different; and a sharp opposition to differentiation or stratification because of biogenetic or socially induced differences in behavior.

The political philosopher Michael Walzer best sums up this dominant mood in social science when he proposed a simple, effective way to achieve equality; namely, to eliminate the ability of some people to dominate their fellows. His point is not to introduce feasibility statements on how this can be brought about or the sort of regime it would take to enforce such a canon or the sort of national conformity needed to implement such a condition, but rather that this goal, this impulse to fairness, that is, "rule without domination," has been central to twentieth-century notions of moral education.[21] But increasingly, a certain discomfort, even disquiet, has penetrated this former condition of near unanimity, this chorus of be-liefs in equality—rewards without costs, creativity without craft, citizenship without responsibility. The core of moral discourse had so shifted from libertarian to egalitarian statements that a certain fear arose that moral commitment would be drowned in an ocean of economic interest groups. And so began the counterrevolution of fundamentalism; again, as a question not of formal philosophy but rather throughout the social and behavioral sciences.

When it is broadly realized that moral education in a post-positivist period of the twentieth century is not delivered by philoso-phers through the front door but by social scientists through the back door, the work of this latter group assumes greater investiga-tive urgency. The moral imperatives of our era—rights against obli-gations; virtues easily articulated over against vices; goods dictated by the marketplace over against evils performed by pernicious planners—all seem to have unraveled. This entire moral climate underwritten by the social-science community crested in the 1960s; a decade in which the Vietnam war was perceived to be an unmixed evil, in which racial justice seemed but a stone's throw away from governmental doorsteps, and in which students were seen to be on the threshold of achieving social parity with much despised admin-istrations and wandering faculties. These simple practical beliefs for a seemingly clearcut social world were intended to guarantee and verify equality over liberty, pluralism over power, individual welfare over competitive markets, alternative life-styles over oppressive straight styles, and happy natives over industrial slaves.

Ethical relativity seemed destined to achieve a total victory in such a social-science *Weltanschauung*. But the relative quickly turned into the absolute: the fear of old "thou shalts" of a theological sort issued into a subtle but no less demanding "thou shalt" of a sociological sort. Benefits were to be pursued without regard to costs, and rights were to be demanded from the state without regard to corresponding obligations to the state. The situation reached a point where guilt vanished and only no-fault forms of punishment and plea bargaining remained. Society turned litigious in this near-perfect atomization of responsibility. It was such a strange climate that social philosophy and social ethics witnessed a rebirth. Nor-mativist political science, phenomenological sociology, biological anthropology, libertarian economics, and instinctualist psychology all began to make comebacks; albeit in more sophisticated intellec-tual packages. Social science, far from foreclosing on the subject of moral education, had only served to disguise and even distort long-standing controversies. A thin patina of data served to hide the same deep well from which empirical researches and moral imperatives both derived.

The area of stratification has been and remains the most sensitive in sociology. This field continues to mirror the larger concerns of American mass society for maximizing equality. Indeed, it is evident, that, in the past sociology anticipated and even stimulated developments in the reduction of racial, sexual, occupational, and educational differentiation. Its pioneer theorists debated the issues while its methodologists created the data base showing the extent to which inequities exist in rank, salaries, roles, and conditions. While an earlier pre–World War II generation sought to minimize any notion that sociology had anything to do with socialism, a latter-day post–World War II generation sought to make quite explicit that the two, however different in outlook and method, did come together in the need for a society in which exaggerated (if not all) forms of inequality and differentiation in all forms of power relations are malevolent and counterproductive and, hence, should be opposed on sociological grounds.[22]

While the accumulation of data and theory did serve to promote a strong sense of egalitarian need, the inability of the former Soviets and even of such democratic welfare societies as Sweden, based on high planning to satisfy the material and the spiritual needs of its citizens, to achieve equitable societies began to cast doubt on a pure theory of egalitarianism or the inevitable, short-term end of stratification. Certain key figures, among them Alex Inkeles,[23] Seymour Martin Lipset,[24] and Paul Hollander,[25] began to emphasize that perfect equality would not only result in minimizing the capacity of a stifling political bureaucracy to implement its credo, but, worse, it would stimulate a breakdown of economic growth, a weakening of technical innovation, and the serious inhibition of personal initiative. The political breeding ground in search of moral man led only to political oppression, fanaticism, and new forms of stratification based upon party elites and political indoctrination.[26] The requirements of a competitive international environment, no less than the demands of those who worked hard and produced much for precisely the sort of differentials which liberal sociological premises formerly villified, led to a broad-scale rethinking of egalitarian premises.

The sociology of stratification was confronted by certain epistemological problems that it could no longer closet: the most serious theoretical dilemma being that issues of stratification have the capacity for, even the inevitability of, infinite regress. Even if one presumes perfect harmony and agreement across racial boundaries, even if sexual differentials in salaries are resolved, the demand structure for further leveling is inexorable: linguistic inequalities, religious boundaries, biological distinctions between the tall and the short, the heavy and the thin, just scratch the surface. What began in the proletarian 1930s as a generation infused with the prospects of marshaling sociology to egalitarian ends, entered the 1990s in a quagmire of anti-libertarian sentiments and a shrill insistence that the state could yet rectify an impossible situation.

The contradictions of supporting repressive statist measures that alone could ensure any enforcement of egalitarian norms became tolerable. Inequalities between the family and the individual, straights and gays, victims and criminals, and old and young all appeared on the scene of sociology, and all were uniformly too complex to be rendered answerable in terms of egalitarian slogans. Intense conflicts between deviant and marginal groups, no less than between such groups and older establishments, became as apparent and insurmountable as earlier struggles between powerful and powerless. Here, too, the weight of social traditions rather than the light-headedness of a social science captured by ideological extemes held sway.[27] Moral instruction in the need for equity only yielded serious practical problems; that is, how to weigh the relative valuational bases of equity and liberty. Solutions seemed ever more remote in theory as the 1980s wore on. The social practice of societies claiming economic equality were revealed to have no such ends and, rather, served only to expand to the political realm the stratification networks; whereas the social practice of societies claiming political democracy had a difficult time explaining, much less dealing with, ongoing forms of exploitation and inequality.

The rebellion against egalitarian modalities in an area such as criminology, in which the very notion of a crime was effectively done away with, in theory at least, through the powerful juxtaposition of crimes and their victims, resulted in a whole new category of sociology called victimology. Everything from the behavior of the victim to compensation patterns decreed by law came to be reconsidered.[28] In this sense, the revolt against "the new criminology," like the revolt against social stratification as such, is a rebellion against taking for granted the routinization of crime. It is, furthermore, an insistence that victims no less than criminals should be treated in terms of the same background variables and rehabilitation potentials. Far from seeing crime as a form of social control, victimology has restored to the literature a notion of crime as a form of social disorganization and moral destruction.

In what must be regarded as a serious departure from conventional criminology, victimology tends to view the innocent recipients of crime as a special repository of the social goods of a moral society. The area of victimology has once again displayed the limits of moral relativism—by perceiving the crime and the criminals as an unmitigated social evil, and the victims of crimes and criminals as the carriers of essential and positive social values. But in doing so, in this intellectual restoration, victimology reintroduces standards of conduct that were thought to be obsolete. The relative nature of criminals and victims comes to be seen as a sharply bifurcated domain between evildoers (criminals) and do-gooders (victims). As the scientific issues have become more muddled by new realities, the sociological reintroduction of normative criteria for public consideration has become widespread.

The areas of abnormal psychology, psychiatry, and psychoanalysis are sufficiently intertwined so that it is appropriate simply to note that they evolved from an early absolutist and reductionist emphasis on wide-ranging categories of abnormality, to a steady narrowing down of the conceptual framework, to the point where abnormality itself came to be conceptually suspect, virtually drowned in a sea of alternative life-styles. As in anthropology, the nurture–nature issue tended to blur the significance of the hormonal basis of masculine and feminine motivational tendencies. At a time of sharply declining moral constraints on individual behavior, neither sociological, neurological, nor psychoanalytic explanations have been especially effective. Deviance itself has been displaced by difference. The assumption that homosexuality, for example, falls into areas of pathology, came to be vigorously denied and organizationally repudiated.[29] Homosexuality, in particular, became a social question and even a political crusade; and in this way it passed from personal trauma to public rights.

A broad-ranging essay cannot possibly resolve technical issues hotly contested within each social-science discipline. However, it is important to note that as social-science discussions on deviance became more tolerant, the demands of the deviant community escalated—for equal rights, for empathy, for open practice, and finally for social affirmation of many forms of deviant behavior formerly viewed as socially and psychologically horrendous.[30] It was only at this point, in the late 1980s, that significant sectors of the social and medical sciences felt moved to reintroduce the place of different types of deviant behavior, to make distinction between sociopathic and psychopathic behavior, and to consider again the place of pathology, both personal and social, in opting for alternative life-styles.[31] Whatever the specific merits of this labeling of deviance, it is evident that the pure empiricism of those who earlier argued for (or against) a discovery of deviance has broken down, to be replaced by a rather frank acknowledgement that moral criteria are both relevant and germane to the scientific analysis of the empirical status of social deviance or personal aggression.[32] The status of norms and norm breaking has thus come to figure prominently in the current revolt against relativism in areas impinging upon personal behavior and even individual life-styles. If the earlier patterns of the medicalization of deviance served as strong, negative moral sanctions, the later socialization of deviance implied equally powerful positive moral supports. That social science is now moving beyond these earlier mechanical formulations is of great significance; the consequence of which is only now an open intellectual agenda.

Political science has created its own equilibrium model, one that derives from Montesquieu and Locke and has stood the field in good stead for most of the century. The grand scenario involves a belief in the plausibility of checks and balances, a society of laws and not of men, a politics of involvement from publics and responsibility from elites. It is not that the best minds in the field of political research failed to appreciate the seamier side of the political process, rather it was their unyielding belief that the system itself was largely intact and that only subsystems, like urban wards, were in disarray.

Voices from the radical left, increasingly insistent on power rather than authority, have been heard; while others from a conservative right, for whom this inherited equilibrium denied the role of norms and values in making the political system work, have likewise been raised in protest. A veritable crescendo of voices was raised concerning the emergence of interest- and pressure-group politics which are only remotely related to the basic goals of the democratic political process. Scholars as diverse as Marshall Berman,[33] Richard Flathman,[34] and John G. A. Pocock[35] have speculated on a political environment in which rights go unquestioned, even assumed, whereas obligations are routinely denied an equal place in any serious analytic scheme of things political. Indeed, even the grounds for a belief in the posssibility of political science have been denied on the basis of natural law and practical politics alike.[36]

The limits of the pluralistic model were reached when, far from a polity in which everyone counts as one, recognition occurred that it takes a veritable economic fortune even to run for minor public office. The stakes in a society differ with the occupational and fiscal place of the individual in that society. What this commonplace fact of life does to a pluralistic model, in which equality is presumed rather than proven, becomes a core issue for political science. Beyond that, declining participation in electoral politics, coupled with a sharp reduction in party commitment, also raises doubts that the older equilibrium model could be sustained in a society in which the assumptions of representative government are subject to criticism and doubt. While some have seen this as a need to improve government by enlarging participation to formerly disenfranchised groups,[37] others saw this same situation as proof positive that populist politics creates the seeds of mediocrity, of popularity contests conducted in the media that displace sounder principles of professional politics.[38] Whatever the implications drawn, it has become evident that the relativism, pluralism, liberalism, and populism of established political-science doctrines have come under intense peer review. As a result, fundamental moral imperatives thought to derive from these doctrinal verities have also been subject to reexamination.

Anthropology is perhaps the last of the major social-science disciplines to discover that its moral tenets are not carved in stone. Buffeted about for nearly half a century between critics on the right, who saw anthropology as an assault on industrial civilization, and by those on the left, who saw the same discipline as a veritable intellectual cockpit of cultural imperialism inside underdeveloped areas, the field itself remained strangely immune from the criticisms of both. Since such critiques were rarely done inside the professional bastions, they were easily perceived as attacks on professionalism rather than science. The typical field reports inevitably emphasize the gentle nature of native peoples, the absence of strife and turmoil across sexual bounds, and romanticized broad continuities between biological and cultural formations.[39] This certainly typifies the sort of field reports issued over the years by the American Museum of Natural History and conforms to the general liberal and relativist positions of fathers and sons and mothers and daughters of the discipline of anthropology.

That this position is now threatening to unravel, if it has not done so already, is made clear in the work of Derek Freeman, who confronts the writings of Margaret Mead on Samoa in particular and the sociological prejudices of cultural anthropology and its forerunners, like W. F. Ogburn, in general.[40] He asserts that the Samoans are intensely competitive; have high incidence of deviance in the form of homicide, assault, and rape; exhibit child-rearing of such authoritarian sorts as to result in a wide range of psychological disturbances including suicide, hysteria, and jealousy; and in lovemaking suffer from the cult of female virginity carried further than Western societies. This raises issues far beyond those of the nurture–nature controversies of the 1920s or of improper research designs; rather, it represents a frontal assault on the moral backbone, or lack thereof, of cultural anthropology. Freeman's work, along with that of Ernst Mayer[41] and Nickolaas Tinbergen,[42] is an attack on the "doctrinal baggage" of cultural relativism and technological determinism, unrefined by authentic observation or realistic evaluation. The ethologist must be in the forefront to restore traditional values and overcome the excesses (pollution and malsocialization) of the modern world. It is from such critical stuff that moral education is refined and restructured. For if the "universals" of ethology and biology point not to any wide disparity between "primitives" and "moderns" but to their close approximation, then the normative bases of ethical judgment must also be seen as a potential tool of cross-cultural analysis.

The science of economics, its practitioners far from being unconcerned about its moral status, probably has been closer to the marrow of moral doctrine than any other social science—if for no other reason than that its founders (Adam Smith, David Ricardo, Karl Marx, Alfred Marshall, and Joseph Schumpeter) all raised problems about the ethical claims of the capitalist market system. Behind its hard-boiled talk of equilibrium models and analysis of social program costs, there remains a strong moral imperative for the well-being of individuals; specifically, how individuals are to be made secure in a world with sharply different social classes (wage earners and owners, factory hands and industrial titans, welfare recipients and foundation directors). Smoothing out the rough edges of difference seemed to be the essential task of the Keynesian revolution as well. Trade-offs between growth and satisfaction worked well enough in periods of sharp growth in a gross national product. But when the growth curve flattened out in the 1980s and real wealth became a constant, when all that rotated was the division rather than the size of the economic pie, then who gets what became a much harsher issue for economists no less than the economy.[43]

From social security to taxation of stocks and property, specific economic concerns began to tear at the fabric of pump-priming solutions, based on high taxation as an acceptable form of benign expropriation. No longer were wealthy classes willing to give up an increasingly larger portion of their profits to the masses to ensure social tranquility. The differences between haves and have-nots resurfaced, much to the chagrin of Keynesians who thought that governmental policies of allocation did away with such angular social disparities once and for all. If Marxians were delighted at this evident collapse of the Keynesian paradigm, it was a mixed and muted pleasure; since socialist economies fared even worse in the wealth reallocation process. The latter-day Marxian saints wrongly assumed that centralization and planning would be resolved by radical reformist sectors. By their own standards, they proved tragically unequal to the task.[44]

Economists have turned away in droves from welfare programs providing "floors" and toward an increasing willingness to let the market create its own floors and ceilings. Reforms centered upon eliminating fiscal waste in management, reducing the size of the federal bureaucracy, eliminating bureaucratic interference in the natural ebb and flow of economies; in general, a return to the nasty and brutish economic environment in which equality would be an outcome and not an input—a universe that permits everyone to finish the race more or less successfully, without any tampering with the start of the race, that is, with the natural or acquired advantages of the individuals running. A kind of neo-Darwinian presumption crept into economic analysis; while at the allocation level *laissez-faire* doctrines of noninterference and nonintervention with the market became primary. From Ludwig von Mises[45] to George Stigler,[46] it became clear that old economic formulas guaranteed democracy only at the cost of dooming or sharply curbing a free-market economy. A growing number of conservative economists, who were willing to run the political risks of subverting democracy rather than the economy, were in diametrical opposition to liberal economists, who continued to be more concerned with floor and ceiling questions.

What this suggests is not the venality or narrow-mindedness of social scientists of the present versus the expansive big-heartedness of social scientists of the mid-1960s. Rather, issues of social systems and moral order are now viewed as more complicated and less susceptible to quick fixes than was earlier thought to be the case.[47] This translated into a review of moral premises upon which our society is grounded. In the world of trade-offs, decisions between alternative concepts of the good, changing standards of what constitutes virtuous behavior, different standards of the true and the beautiful, the social sciences were compelled to redirect their energies—and take much less for granted in the way of the malleability of norms than they were formerly prone to do.

This brief sociology of knowledge review of the social-science literature suggests that issues of moral education are not fought out in classes on moral education but rather in research designs, experiments conducted, and theories dreamed up by social scientists. It is now apparent that the ghost in the social-science machine is ethics. At the start and conclusion of every major piece of empirical research, or theoretical paradigm developed by social scientists, has been some driving moral imperative about personal rights, racial balance, or sexual equality.[48] But this moral dimension has been obscured by those for whom facts speak for themselves. Along with the collapse of positivism in philosophy, functionalism has likewise suffered badly as the only available social-research design.

It is better to view social scientists as sensitizing agents in moral discourse than as engineers of the soul capable of resolving inherited ethical issues. They have not been able to do so in the past and probably will fare little better in decades to come. It is hard to know whether changing moral climates compel a rethinking of social-science premises or whether new discoveries in the social sciences compel a changing attitude toward particular ethical commitments and beliefs. In a sense, this is a less important distinction, less important than meets the eye. Casual priorities may be important to those who wish to claim a pedagogic advantage, but to the citizenry at large, such matters of causation are entirely derivative and secondary. The main issue is now clear: how do the social sciences smuggle moral imperatives into their findings? At a technical level, the question becomes: how does the subject matter of moral education serve to sensitize social scientists to their own premises?

The ability to get beyond an era of a ready-to-wear assumption that social science and morality are irrelevant to one another, or at least independent of each other in development, itself marks a large step forward and upward. If this involves a return to notions of political morality no less than political economy, then so be it. This is a far more eloquent outcome than continuing to presume that moral discourse can seriously take place without relevance to the empirical findings of social science or, just as tragically, that social-science research is devoid of value contentions or value intentions. One no longer hears the positivistic claims that moral education is an impediment to research as such. Both dualist and positivist premises are reductionist, dangerous, myopic. Exclusivity of moral claims results in vague linguistic analysis that lacks the shank of reality; whereas exclusivity of social science results in a constant repetition of past mistakes, a hubris of concreteness that requires constant revision and periodic adjustment like the Ptolemaic calendar. It can be expected that this dawning recognition of the need for cross-fertilization of social science and moral awareness can preserve our intellectual classes from the most serious consequences of spiritualism and reductionism alike.

Abraham Edel, in his search for a "common ethic," located in the theory of the valuational base as the source of that naturalistic ethic, makes an appeal to the social sciences: "We are, in short, looking to the full scientific perspective which embraces the lessons of the human sciences from biology to history, and applies them to the contemporary life of society and the individual, to help us fashion a general outlook on our world and ourselves."[49] While this ambition remains *prima facie* worthwhile, the problem remains that the social sciences, far from offering any sort of unified picture from which ethical guidelines can be drawn, are, in fact, deeply mired in contentious, parochial claims that are as old as classical discourse in philosophy.

This is by no means to conclude on a pessimistic note or to assert that ethical claims are more profoundly stated in absolute terms, without reference to contemporary social-scientific research. However, it is to make clear that no heavenly mandates or cookbook recipes can be offered in a scientific world in which ethical presuppositions are constantly being reworked under highly pressurized conditions. This aspect of a historically grounded ethical framework, one that takes normative propositions as themselves empirical data, is very much on target. Edel begins the process of synthesis opened up one hundred years ago by Durkheim's call for a moral education based on sociological premises. But to do this big job effectively requires the full cooperation of all the human sciences, without *a priori* imperial presumptions as to which one of them has the answers. In the past, ethical guidelines were smuggled into social research as an *ad hoc* demonstration that one or another of the social sciences was supreme. In the future, such side shows will yield to the main show: how to ensure that moral discourse is incorporated into, and is in turn informed by, the full range of social research.

15

Social Science and the Great Tradition

Contrary to what might be imagined, social-science publishing in the United States is not a recent development. Quite the contrary, broadly conceived, it began almost coincidentally with the founding of the American Republic. Specifically, the issuance of the first census-tract information in 1790 represented the introduction of social-science–data publishing. In addition, even before the ratification of the American Constitution, the issuance of *The Federalist Papers* may be viewed as the first substantial effort in publicizing the politics of the "first new nation."[1]

The continuing legacy of demographic information, Constitutionally mandated, thus provides an early example of social-science publishing in the United States. And from that time forward, the United States Government Printing Office has been in the forefront of such activities. To be sure, much of this has been offered to the public as raw data and statistics rather than social science as a controversial or a critical activity. But given the variables covered in statistical data bases—everything from racial and religious composition of the population, the size and geographical distribution of the work force, and the segmentation of labor markets, to deviance and crime and alcoholism—it is clear that the government has long recognized social science not only as a constituent element of its publishing program but also as one of the elementary requirements of a free society: the right of its citizens to know.

From the outset of the Republic, many government special reports, congressional committees, presidential commissions, and legal briefs have had a special impact on the evolution of social science. For unlike its European counterparts, the United States from the beginning established a connection between research findings, social policy, and government activities. The publications efforts of the United States Government Printing Office consolidated early on both the professional and pragmatic areas of social science in America. Everything was disseminated, from reports on corruption in the Indian reservations to child-labor–law abuses. This literary as well as legislative activity led to new rules and legislation in which social-science researchers performed an increasingly central role.[2]

In the nineteenth century, the publication of social-science overviews by the large commercial presses (especially Henry Holt, Boni & Liveright, Macmillan, Scribners, Lippincott, and H. R. Lea) for large public consumption was introduced. Thus, early social scientists—from Alexis de Tocqueville on democracy in America to Harriet Martineau on manners and mores in the United States to Charles Horton Cooley on the social psychology of individuals and groups to James Bryce's two-volume classic on *The American Commonwealth* to William Graham Sumner on mores and customs—were all widely published by commercial houses.

We need to keep in mind that the social sciences were a much smaller entity in sheer numerical terms at the start of the twentieth century than at its end. Social scientists often viewed themselves as public commentators on societal and civilizational issues rather than as narrow specialists in highly parceled and refined areas of research. Indeed, they were part of the "clerisy" of letters imported from England more than any specialist group defined over against people of letters, as seems currently the case. In short, the power of social science was more in profession of beliefs than professionalization of chores.

One of the great myths about social-science publishing is that commercial presses ceased issuing such works in a dim and distant past. In point of fact, if we take into account that firms like the Free Press and Basic Books are subsidiaries of major publishers like Macmillan and Harper/Collins, respectively, we can see that this is not the case. In fact, a strong argument can be made that some of the best titles in the social sciences continue to be issued by commerical firms and by authors who continue to have strong public concerns. And if we take seriously the fact that firms like Oxford University Press are essentially profit-making organizations (and university presses in a rather peripheral sense), the notion that scholarly works are a preserve of tiny university presses can be properly laid to rest.

One must hasten to point out that commercial publishing was, for most of the nineteenth century, elite oriented. Only with the introduction of the "penny novel" toward the close of the century, as a response to the evolution of mass-circulation newspapers, did this situation change.[3] In this, American publishing followed British trends. And if the American social scientists did not become household names or have their works put on the mantle along with Herbert Spencer's, Charles Darwin's, Shakespeare's, and the Bible, then they at least were viewed as organizing the mentality of America during its formative years. Lord Bryce on the politics of the American Commonwealth, William James on psychology and its relation to religion, and Auguste Comte on the religion of sociology were all well-represented on commerical publishing lists.

In addition, the major commerical houses brought out English-language editions of Europeans, including the major writers from Germany and France. Even such dangerous figures as Karl Marx and Frederick Engels were published early on by Richard Henry Dana in the *New York Herald-Tribune.* Indeed, it is only when tendencies toward professionalization and specialization set in that the great gap between popular writing and social-science writing became unbridgeably wide. In such a bifurcated environment, socialist publishing houses, like Charles Kerr, served to bridge the gap between the popular and professional, especially in the areas of economics and politics. These radical and atheistic presses vied with the specifically religious publishers to reach large numbers of people, and they did so in the context of the terms made popular by Andrew White: "the warfare between science and theology." In this sense, social science was a by-product of the struggle between secular and clerical tendencies rather than something advanced in its own right.

Of great import was the development and institutionalization of social work and welfare in America. With the rise in the late nineteenth century of foundations, such as the Russell Sage Foundation, large numbers of monographs and books intended for practitioners were put out. Russell Sage has, in fact, one of the longest continuous series of books in the social sciences in the United States. Such foundations and agencies not only put out practical manuals for social workers but also published fundamental theoretical tracts in the basic sciences, often translations from European sources. Indeed, by publishing Ludwig Gumplowicz's *Foundations of Sociology,* the

Russell Sage Foundation Press served to introduce social-conflict theory to the United States.

Rarely spoken of is the emergence of racial and ethnic publication groups. In this connection, the development of black publishing between the turn of the century and World War I is especially noteworthy. Publishing firms like the Colored Cooperative Publishing Company, J. A. Rogers Publications, Negro Yearbook Publishing Company, and the Association for the Study of Negro Life and History put out a large variety of books that merged fields of social science and social history—a merger, by the way, that is characteristic of the leading figures of black intellectual life throughout the century.[4] *Negro Yearbook* was especially important in that it involved the participation of such sociologists as Monroe N. Work, Emmett J. Scott, and Robert E. Park as editors and entrepreneurs. Such books as Carter G. Woodson's *A Century of Migration*, Eugene McDougle's *Slavery in Kentucky*, and the sociological novel by Pauline Hopkins, *Contending Forces*, published in 1890 and aimed at "developing the men and women who will faithfully portray the innermost thoughts and feelings of the Negro with all the fire and romance which lie dormant in our history," are especially noteworthy.

With the rise of professional societies in various fields of social science, there also arose journal publications. In the twenty-year period between 1890 and 1910, just about every basic social science—sociology, economics, anthropology, psychology, and political science—developed its own publishing program, usually emphasizing journal rather than book publishing. And as specialization set in, these professional groups increased their journal and book offerings to satisfy specialist demands.

But again, with the rise of foundations and professions, the character of social-science publishing changed substantially. A premium was placed on exactitude, on data, and on value neutrality—elements which limited public interest and, hence, reduced commercial interest in the publication of social science. Micro studies displaced macro studies. And society and community replaced civilization. Journalists like Walter Lippmann, Albert Jay Nock, and Irving Babbitt replaced the social-science writers. Popular attitudes were shaped more by H. L. Mencken and *The American Mercury* than by social scientists. In short, the growth of professionalization became a basic cause of the widening gap between popular writing and social-science research.

In such an environment, the emergence of university presses

played an increasingly significant role in the social sciences. Such presses as the Teachers College Press of Columbia University, and the Johns Hopkins University Press began to publish the results of applied researches in the fields of education, law, and medicine. And, in turn, such forms of publication became a bona fide means for professional advancement, especially within universities that were emphasizing graduate programs. Many of these earlier publishing firms were linked to the dissertation process. Since publication was a doctoral requirement (at least until World War II, and in some universities long afterward), university presses were, to use an inelegant word, often involved in "vanity" publishing in order to satisfy the flow of demands of teachers and researchers, not to mention to ensure their own solvency. The author of a thesis was charged for the costs involved in the production process—usually just enough to cover basic costs. The King's Crown imprimatur of Columbia University Press and the Teachers College Press pioneered this form of publication, which opened the way to monographic publication of empirical social research on a large scale.

Up until World War II, university presses tended to publish mainly monographs of a dissertation level or items of a highly specialized sort, whatever their fields of concentration. This corresponded to departmental theses requirements that necessitated the printing of anywhere from fifty to a hundred copies (at the student's expense) as a condition of graduation. The better works, or at least the enduring works, continued to be issued by commercial houses between the two world wars. Gordon Allport's *Social Psychology* was released by Houghton Mifflin, and Harry Elmer Barnes's *The New History and the Social Studies* was issued by Century (the precursor to Appleton-Century & Crofts—Indeed, it was Appleton which released all of Herbert Spencer's major works between 1876 and 1897, including the *Principles of Sociology and Social Statics*). William Graham Sumner's *What the Social Classes Owe to Each Other* was issued by Harper's in 1883, while Robert H. Lowie's *The History of Ethnological Theory* was published by Holt, Rinehart & Winston. In short, the publications in the social sciences that flowed from the commercial presses were substantial in number, many works having large press runs, numbers of editions, and even high quality.

I must also pause to make special mention of the impact of the works of Thorstein Veblen in economics, Walter Lippmann in political science, David Riesman in sociology, Margaret Mead in anthropology, and, more recently, Peter Drucker in business and William

Whyte in organization. Almost like a continuing thread throughout the century, the books of these figures sold in massive numbers to wide and disparate audiences. And even if they were, as a result, marginalized in academic terms and often subject to professional derision and rebuke, they provided a model of expansion in the book world that was not lost as the period of specialization in social research developed after the conclusion of World War II.

It must be noted that the war years themselves served as a sort of catalyst to social science. This was an ideological as well as a military conflict. "Fascism," "communism," and "Nazism" became household words. The need to satisfy this market throughout the decade of the 1940s was very great. Not only writers on these themes, like James Burnham and Dorothy Parker, emerged, but also university-based scholars, often from history, like Henry Steele Commager, came to the fore. In addition, socialism, with a strong predilection for social-scientific as well as philosophic generations, became a renewed buying area. The works of Marx, Engels, Lenin, Trotsky, and, yes, Stalin became staples of intellectual households. The writings of socialists like Eugene Victor Debs and Norman Thomas likewise permeated many American homes.

Too often it is thought that the line from education to profession is straight and narrow. Nothing could be further from the truth in social science. The line was bent into a curve. The impact of the socialists as well as the fascists is all too well known to be discussed here. Suffice it to say that they had in common anti-intellectual credos and anti-Semitic practices, but the role of socialist ideology, particularly in social-science publishing, can hardly be exaggerated. It runs like a thread throughout the first half of the century, dampened only by the perverse nature of totalitarian systems that began to discriminate against sociology and psychiatry as bourgeois tools of imperial powers. It was thus the inner logic of totalitarianism and not the external pressure of a bourgeoisie that finally severed the cord between socialism and sociology (along with other social sciences).

The years between the two great wars of the century, essentially between 1920 and 1940, saw the first specialization take place in social-scientific publishing. With the emergence of major research departments of sociology at the University of Chicago and the University of North Carolina, and similar world-class work in political science at Yale, psychology at Harvard and Columbia, and, above all, economics in all of these schools and in specialized agencies, the university presses became critical outlets for specialized monographs, very often works done by scholars on campus. Hence, it is not an accident that the University of Chicago Press became the key publisher for people like Robert Ezra Park, Florian Znaniecki, Herbert Blumer, Everett Hughes, and William Ogburn; while Yale University Press became the key publisher for Harold Lasswell and Lloyd Warner, among others. In short, with the emergence of social science as a set of specialized tools and agencies, the prospects for market expansion in book publishing also increased.

University presses are, however, set up to serve many academic constituencies, from those interested in the history of French literature to those favoring molecular biology. As a result, such presses provided only a limited outlet, usually to already established "stars" of the social-science world. University presses also did not develop a marketing strategy aimed at social scientists as such. Rather, they tried to reach the intelligent and broad-ranging scholar. As a result, the expanding needs of social research for publication outlets could not adequately be met by university presses, despite their Herculean efforts to serve their universities and their scholars in residence. One might argue with some justification that the university presses continue to service only a small strata of social-scientific authors and readers—in part because the format and function of such presses are less linked to the research process of social sciences and more to the classical humanistic culture of a well-rounded individual being serviced by a well-rounded university press.

After World War II, university presses began to multiply, temporarily satisfying demands of the academic tenure and promotion process no less than scholarship and research. And in such centers of social-science research as the University of Chicago, the presses virtually became social-science presses. Other presses began subspecialization, often in partnership with foundations. Thus, Princeton became the center for psychoanalytical and historical publishing with the support of outside foundations; Yale, likewise, became a center of political-science publishing. European houses, such as Oxford and Cambridge, developed important outlets in the United States, often setting up independent editorial and marketing units. The more popular of social-science authors were published in substantial editions. David Riesman was published by Yale University, C. Wright Mills by Oxford University Press, William F. Whyte by the University of Chicago Press. Thus, the university and scholarly presses began to satisfy the same general-market needs that were serviced at the turn of the century by the commercial houses.

Commercial publishers like Doubleday, Macmillan, and Random House renewed and increased their interests in social scientists as popular commentators. This was accelerated by the rise of the "paperback revolution" in the 1950s. Harper Torchbooks, Meridian Books, and Doubleday Anchor turned to the social sciences with a vengeance to fill out their lists and satisfy a mass interest in "the American century," especially the social psychology of honored Americans.[5] It might well be argued that the mass paperback and the popularization of social science were closely linked.

After World War II, with the explosion of academic life and academic social science—it should be noted that college and university enrollments went from roughly one million in 1940 to over fifteen million in 1980 (and that does not include the junior college population)—it became clear that social-science publication by university presses or foundations would simply prove inadequate to meet the expanding needs. Into this environment came an innovative development—social-science publishing as an independent activity.

Among the early pioneers of this were Jeremiah Kaplan, who started the Free Press as a civil-libertarian social-science press in the late 1940s, and, at roughly the same time, Frederick A. Praeger, who began, or rather retooled, Praeger Publishers (which had been a European press that was forced to leave Europe under Nazism). The disciplinary emphasis of the two houses was quite different. The Free Press aimed to satisfy a new specialist classroom and text audience at the top of the market, while Praeger published the dissertation literature in the social sciences. The Free Press and Praeger were highly successful in fiscal terms, and this opened the way to a general social-science publishing effort targeted at the social sciences themselves.

And when one speaks of a publishing effort directed at the social sciences, this in itself represents the great divide between all earlier publishing efforts in the social sciences and the postwar environment which took as a given the social sciences as a unique and identifiable market. The social scientists were readily identifiable as a group and in select areas—a huge plus for publishers. One could reach them as a group for relatively modest sums and be certain of a receptive market. In this explosion of social science, there was an implosion in place; that is to say, growth was in select and identifiable environments—and practitioners had a relative affluence which permitted book buying to take place.

The recent change in social-science publishing, from commercial to scholarly presses, took place as a result of parallel events: the

growth of specialization in the social sciences that resulted in frac-
tured, smaller markets and, hence, smaller unit sales; and the corresponding and opposite tendency toward monopolization in commercial publishing that resulted in the need for ever larger press runs to
justify acceptance of manuscripts. Only at that point, with the
"merger manias" of the last thirty years, did noticeably different
patterns arise. But the movement of social-science publishing from
commercial to university-based activities is a separate and distinct
subject, quite removed from the history of social-science publication
as such.

Academic Publishers in psychology, Westview and Ballinger in
political science, Transaction and Sage in sociology, and Basic Books
in economics and policy were just some of the new houses that
emerged, with relative degrees of consciousness to satisfy a social-
science marketplace that itself had grown to a half million souls.
The European firms of Elsevier, Pergamon, and Kluwer, among
others, have specialized in the applied areas of social research and
information technology. The emergence of *Psychology Today, Demographics, Human Behavior, Human Nature,* and *Society* made it
clear that the social sciences were ready to move beyond their
cocoon-like emphasis on professional needs and to again seek out
the larger world of intelligent laymen. But this effort has produced
ambiguous results; journalists have become wiser in the handling of
data, social scientists have become more precise and cautious in
their foci, and a general suspicion has emerged, fostered in part by
the social scientists themselves, that they are not neutral observers
painting on a blank canvas but active partisans of causes, interests,
and policies that have no more and no less a scientific warrant than
those agenda items put forth by other segments of the public.

As a consequence, not all ventures in social-science journals and
serials have been successful, any more than all book enterprises have
been. Many of these ventures have been ill-suited to specialized
environments; others have misread the size and strength of the marketplace at any given time; still others have failed for want of creative editors who could cross over from technical to popular forms of
communication. Yet, some have made the transition. Publications
like *The Public Interest, Transaction/Society,* and *Psychology Today* all show the potential in this area. Further, with such scientific
publications as *Scientific American* and *Science* (the weekly produced by the American Association for the Advancement of Science), the social sciences have established a niche as part of the
broader scientific community.

Specialized serial publications have become a major force in the social sciences, as they have in the biological sciences. There are now specialized publishers in serial publishing catering almost exclusively to library and archival needs. In short, over the century social science has moved from an aspect of the book world to a central place in the information and knowledge explosion. It is precisely this transformation, from "literary" to "informational," that has been the most dramatic change in the history of social-science publishing.

Along with the benefits from niche publishing came penalties. The basic problem is that the new specialized publishing environment encourages fragmentation and atomization. The earlier period may have exhibited technical weaknesses, but at least the reader was provided with a sense of the whole and with relevant contexts aiming to make "sense" out of big problems and paradoxes. Thus, the publishing condition serves to reinforce those positivist tendencies that already threaten to leave interpretation of large-scale events to non-social scientists. But such issues are better left for other papers and other times.

What is clear is that social-science publishing is not just a small part of the history of social science as such. The world of publications well illustrates the major themes with which the area of social science is still struggling: professional versus popular emphasis, theoretical versus pragmatic orientations, and impressionistic versus technical visions of the social environment. These polarities continue to coexist. Indeed, the publishing arms just described tend to serve a wide variety of visions and views—not always in a harmonious fashion.

As the twentieth century has progressed, it has become clear that the fates and fortunes of two areas—publishing and research—have become increasingly intertwined, with the emergence of a professional cadre of people who are schooled in academic publishing, archival retrieval, and in social-science research techniques. This has made for a better educated public and, surely, a better-balanced social-science environment.

The real issue, and one that beckons to bedevil us for the remainder of the century and beyond, is the extent to which a sense of the democratic order that has sustained social science from the birth of the "first new nation" has been lost in the process of technification. Does the sponsored-research report intended for the eyes of a select few displace the writings of free spirits intended to be read and reviewed by the many? Can these varieties of literary expression peacefully and creatively coincide in the marketplace of ideas? But these concerns clearly spill over into larger considerations of the relationship between information and knowledge—and that goes beyond the confines of this purely historical examination for maintaining social science.

Reconstructing social science involves reconfiguring those factors that define the nature of the inner-working life of social science. By treating the area of publication as integral to the definition of professional fields rather than as a simple mechanism for getting information or ideas from creator to end user, we arrive at a cutting edge of social science as such. For it is the entire cycle of knowledge and its dissemination that permits a wider, more ample vision of social research as self-reflexive; that is, as a wide-ranging entity which involves everything from the sale of books or journals as commodities to the promulgation of ideas therein contained for the purpose of changing everything from public policy to public opinion. In such an examination, we can begin to see how the cumulative nature of social-scientific findings and results go considerably beyond the search for individual fame, into the very sinews of public value. One can anticipate this process becoming even more pronounced in the century ahead—especially as aspirations for democratic life link up with demands for better data and clearer thinking to assist in the realization of such aspirations.

16

Social Science as the Third Culture

Discussions about the rights and wrongs of higher education are so daunting in number, if not in solutions, that it is probably best to draw upon one's own experiences, at least as a starting point. It is not that experience is a better guide than reason in human affairs; rather, the ground of reason can best be established by a sense of shared needs, wants, and traditions of the individuals, administrators, teachers, and students who coexist in this strange world of higher learning. What follows, then, is a set of personal observations rather than a scholarly treatise or, for that matter, a policy wish list.

In an earlier effort,[1] I attempted a new critique of "The Two Cultures" theme, only to conclude that whatever the problems existing in the competition between science and the humanities were not to be resolved by some mechanical transfusion of one into the other in the name of an integration lacking all specificity. Indeed, however intellectually disquieting "dualisms" and their tendencies to reify the world have been, such dualisms at least have had the merit of being prima facie accurate. Take as an example professionalism. In looking at the realities of this concept and moving behind the current debates on the subject, it turns out that professionalism is not an escape from reality but an effort by the scholar to avoid ideological positioning. This is admittedly easier said than done. Indeed, professionalism has itself been labeled an ideology—which only leads to a strange situation wherein the very act of seeking a scientific standpoint becomes suspect or worse.

While one can argue that distinctions between the philosophic and the scientific (or more readily the distinction between ethical and empirical matters) can be removed by considering all the world as an ideological stage, such a synthesis is clearly worse than the problem. It leads to a critique of our common heritage as Eurocentric, ethnocentric, and similar fashionable radicalisms. By the same token, to consider forms of knowledge as matters of syntax and language protocols is likewise a disastrous form of positivism; one that makes the character of knowing subject to principles of logic without regard to the actual divergencies in the world. In the closing decade of the twentieth century, it is evident that dialectical materialism and logical positivism are spent doctrines, emptied of innovative frames. Thus, it is that the Peircian "fixation of belief" between scientific and authoritarian modes continues in full force; nowhere more so than in educational systems and policies throughout the world.

The problem is that in such practical recognition, there is a tendency to insist upon universalist standards when it comes to the scientific and to deny holistic premises in the name of partisanship when it comes to the humanistic arts and the "soft" social sciences. This dualism serves, in effect, to enshrine the most bizarre sort of epistemology, one that sees the developing world as needing exact scientific standards for industrial evolution and inexact ideological formations in the conduct of politics. The results are societies that demand scientific advancement in the name of ideological goals and a collapse of nerve on the part of policymakers and social researchers and higher educators when it comes to the meaning and nature of development. In this way, everything from state-sponsored mass terror to unsupported claims of grandeur come to characterize the development process in Third World nations.

Caught in a vice-like grip of a reductionist scientism on one hand and an overblown metaphysic of partisanship on the other, such Third World nations, far from resolving the dualism of the advanced nations, through their higher educational systems only served to institutionalize a far worse and less efficient form of dualism. In nation after nation in Asia, Latin America, and Africa, this terrible schism between scientific technology and humanist ideology is not only accepted but also preached by foreign dignitaries, who in the name of solidarity manage a form of patronizing paternalism of unmatched dimensions. It usually takes the form of preaching a doctrine that the ruthless authoritarianisms practiced in many portions of the Third World are somehow a form of "higher" democracy. In this way, they remove any semblance of universal meaning formerly attached to the concept of freedom, liberty and democracy are emptied of content, and people are left to choose between utopian rubbish and cynical denial of the political process as such.

This may be a rather elaborate way of beginning to address the role of universities in developing areas, but if you bear with me, I hope to make this linkage of the personal and the professional clear. To start with, rather than engaging in woes and moaning about the U.S. Department of Education data—showing that 37 percent of students who graduate from college do so without a course in history; 45 percent without a course in English literature; 65 percent without a course in philosophy; 77 percent without a foreign language—I want to address the converse issue: what about the problems of the 63 percent who *do* take courses in history; the 55 percent who *have* courses in English literature; the 35 percent who take a philosophy course; and the 23 percent who study a foreign language. To ask this question is to move beyond the shock value of quantitative recitations of illiteracy into qualitative varieties of ideology. This is by no means intended to slight the need for vast improvements in basic curriculum matters but rather to address the contents of education that are too easily fudged by easy outrage at some lopsided data demonstrating cultural illiteracy.

The issues that face those subject to education are similar the world over. It is a point of my remarks to show that while numbers change, and circumstances in higher education differ from nation to nation, certain issues in moral discourse and legal philosophy are universal and are subject to a general analysis; while other disciplines, such as physics, chemistry, and biology, are specific to historical eras and geographical places. Hence, it is best to conceive of the tasks of higher education in social-development rather than national-interest terms.

The delicate balance between agreement on the ground rules of debate and discourse amidst differences on specific intellectual choices offers a way of examining the big questions about the uses of social science. Neither scientific nor humanistic studies offer an ideological monopoly. Schools may be private, sponsorship may be regional, but social and cultural formations are public and universal. They are not owned by a particular sect or creed. Indeed, in its nature, validity in the world of learning is subject to multiple scrutinies. Yet, it is precisely such relatively simple assertions that seem hardest to implement.

For while the claims of science are viewed as universal and subject to rules of evidence and experience, the claims of the social sciences are seen as particular and subject to partisanship. Indeed, rather than being viewed as shortcomings in the teaching of subjects ranging from anthropology to sociology, partisanship and emotionalism are frequently held as prerequisite for proper thought and their absence as sound reason to mistrust the bearers of objectivity in human affairs.

The major theoretical conclusion becomes axiomatic: when talking about the status of the social sciences, we are, in fact, concerned with which cultural formations are to be emphasized. How much Vilfredo Pareto in place of Émile Durkheim? How much Chinua Achebe in place of William Shakespeare? How much John Coltrane instead of Wolfgang Mozart? And how much Alvin Ailey Dance Troupe instead of the Bolshoi Ballet? Such debates are dissimilar to discussions among physicists over the degree to which Newton's science of mechanics is replaced by Einstein's rules of relativity. While recognizing the distinction between rules of evidence in science and strategies of discourse in social research, there is, nonetheless, a broad consensus about quality, or at least about the pantheon of greats who provide rewards for study. In this sense, the choice between Pareto and Durkheim may be less significant than the recognition that both belong in a sociology curriculum, while neither merits official blessing. To be sure, the current rejection of the official doctrines of Marxism–Leninism in the former communist world indicates the repugnance by those who have lived lifetimes under totalitarianism felt about official thought. At the root of things is a choice between fitting the facts to the theories, and fitting the theories to the facts. Ideology insinuates the former, while science declares in favor of the latter framework.

The social sciences have been deeply impacted by current levels of technology.[2] First, there are the hyphenated disciplines ranging from political economy to social psychology; second, there are fields that overlap all the social sciences, from probability theory to technical training in computer concepts, automation, and information distribution operations; third, there are deep changes in the mix between general theory and specific application. In a sense, then, this issue of universality and particularity is itself a red herring. It is clearly the case that even the greatest and most widely recognized masterpieces are rooted in a specific times and places, whereas the most localized of cultural formations view themselves as having the potential for universal appeal. The struggle for a folk art is therefore far more than a claim on a tiny segment of practitioners. It is an

urging, sometimes even an insistence, that folk art contains the
same grains of universal value and appeal as so-called high culture,
plus the presumed joys of patriotism. It is precisely the admixture of
power and culture that forms the essential Lasswellian question
of what gets taught, by whom, and just how. And the examination of
this admixture is itself a task of social research.

The problem of education and development becomes complex at
intimate levels of the relation between power and principle. Within
more specifically defined areas in space and time, there are subtle
choices. In theater, for example, how much Eugene O'Neil vis-à-vis
a Harold Pinter; or in music, how much Mozart with respect to how
much Prokofiev; or in theories of causality, how to weigh the naturalism of a David Bohm against the empiricism of a Niels Bohr or an
Erwin Shroedinger. So, at a second level of education, the task is not
one of investment, either public or individual, between scientific
work and humanities as such but what particular strategies are
adopted to implement certain doctrines or promote certain interests.

Each national educational system must also consider the issue of
choice. As a consequence, curriculum decisions link up with cultural climates to form yet a third level of intimacy with the subject
at hand: *which* Prokofiev?—the early Russian prodigy, the Parisian
exile, or the dedicated, if sometimes defamed, Soviet nationalist? Or,
in sociology are we to emphasize the work of Durkheim on the
division of labor, on suicide, or on (his later writings) moral education. Each emphasis will lead to different intellectual outcomes and,
more important, pedagogical consequences. In short, social science
is neither the exclusive property of an ideology nor the plaything of
technicians. It is properly what the struggle for the minds of men and
women is all about. That is to say, the struggle within each scientific
discipline is no less an ideological forum. Such struggles take place
within each cultural or liberal-arts segment of university life. To
recognize this situation is not the same as advocating a particular
solution; nor does recognition of the ideological component doom
the quest for higher learning to base considerations. Indeed, the
standpoint of the sociology of knowledge is but an additional, albeit
critical, element in the search for universals.

Even this three-step notion of probing the world of learning is a
vastly over-simplified way of viewing the problem of social-science
aims in contemporary higher education. The struggle of ideology
takes place within the cultural formations and layers of a society.
That is to say, the struggle between Plato and Aristotle, the differences between Kant and Hegel, or Marx and Weber, are part of the

common wisdom of their respective ages and must be understood in ours. For what is at risk in scientific research is not an end to knowledge but to the moral or intellectual dialogue that the classical splits in science and the arts has represented for past generations.

Higher education is presently subverted as much by a sense that the tradition and canons of Western thought constitute building blocks of civilization as by a categorical denial that Western civilization even exists. Without a dialogue among the classics of the social sciences, no less than the classics as such, there can be no culture of civility—only the incivilities of the dogmatists and fanatics. A new extremism has come into being made of deconstructionists; new historicists; people in gender studies, ethnic studies, and media studies; and a few leftover communists and fascists. These loud remnants would like to use the social sciences as staging areas for political action. It is clear that this tendency is scarcely confined to any single discipline or field. Indeed, as the political, economic, and social bankruptcy of the Second, or communist, World becomes increasingly manifest, the critics of Western-style democracies shift their affections to the presumably less vulnerable nations of the Third World. But if the developmental process has been so thoroughly stymied in nations of Europe, which once exhibited high levels of industrial achievement, what are the nations of the Third World to expect by adopting ideologies of a dogmatic sort without even having had that previous experience with the engines of economic change? To ask the question is in this case to answer it: not much.

Sidney Hook, just prior to his death, wrote a paper on "The Politics of Curriculum Building."[3] In it, he pointed out that the astonishing thing about those with revolutionary allegiances "is the degree to which those who pretend to or profess a Marxist orientation interpret Marxism not as a scholarly scientific approach, but as a commitment to a variety of political causes." The essence of such persuasions is purely tactical: opposition to American foreign policy whenever conflict is involved with either a developing area or socialist-bloc nation. As a result, precise analysis yields to abstract condemnation as a way of academic life.

At the other end of this building-block approach to learning is the traditional conservative standpoint, in which the canons of learning are unshakable and the classics handed down from generation to generation in an unchanging way. Veneration substitutes for analysis. And the idea of ethical knowledge reduces to dogmatic assertions that historical progress consists solely in imbibing the inherited wisdom of a specific discipline; while the real-world events are seen only as a series of footnotes to these figures (when right) or deviations from these figures (when wrong).

In this sense, the challenge to developing a serious pedagogical agenda, one that would incorporate the scientific and technological revolutions of our time, is twofold. First, it would by convention be necessary to conduct research with the belief that the history of ideas proceeds from one great figure to another—as if minds are not rooted in place, space, and, above all, in the human form. Implicit in this belief is the sly view that morals are unchanging, that they need only be imbibed once and for all time and need not yield to changing characteristics of economic development writ large. Second, one would have to hold with dogmatism to the figures who are deemed inside or outside the pale of authorized and appropriate figures or concepts worthy of discussion. In this way, the canons of a field replace field observations.

To enhance social science means to argue about what is living and what is dead constantly and fearlessly. This mediating function is done through the social testing of propositions as to what constitutes goodness, truth, beauty, tragedy, and comedy. To struggle for a humane social-science tradition is also to use social science as such, to distinguish what is particular to one age from what is universal for all ages. It is not a religious *sin* that three-fourths of American students do not take foreign language courses, but it is a political tragedy that three-fourths of American students cannot communicate with nine-tenths of the human race living beyond the pale of the English language. Likewise, it is also not a catastrophe to assert English as the universal language of science in the twentieth century—as German was in the nineteenth century, as French was in the eighteenth century, and as Latin was throughout the medieval world, despite the fall of the Empire as such. To confuse the need for broad commonality of language with acquiescence to presumed imperial sources of power is only to postpone and, in certain instances, to doom the process of development, not to enhance independence or advance standards of living. Illustrative of this distinction is the decision, after thirty years of suppression of the study of English—as a colonialist remnant—in Sri Lanka, to completely reverse the policy. English is now mandatory as the essential mode for again becoming part of international development.

Social-science research does not have a monopoly on those given to nostalgia. It is, rather, the sum total of usable thought derived from the recent past that is of service to us in present. Fashion in social-science disciplines changes; and that is also why the struggle for liberal arts has vitality and worth. The social sciences are not a fortress against development but the cultural system that has come to eminence in the twentieth century that can fuel changes and make them less jagged, less severe in their human consequences. No one doubted the authenticity of development in the Soviet Union after the destruction of the czarist system. But one has every right to doubt the need for the destruction of sixty-million Russian souls to achieve such developmental goals.[4] Beyond that, if social science does have a humane dimension, it is to show what options, if any, are available to brutal first choices. Perhaps the preservation, rather than the destruction, of social-science learning under Stalinism might have provided just such a cushion against the madnesses that took place between 1932 and 1952. Revolutions bring about exaggerations and mannerisms in which a challenge of contemporary culture is issued to replace the inherited culture. In such circumstances, it is the task of the university and its social-science components, to sort out such challenges. In doing so, the core of education is itself enhanced without being subject to ossification through tradition on the right or to fanaticism through futurology on the left.

A simple illustration may suffice in explicating this point: the work of such contemporary figures as Harold D. Lasswell or Robert K. Merton may simply vanish from view. But the more likely outcome in the long tradition is that such figures are retained in a select way. Issues of personality are examined in terms of more classic views of power relations; the role of dysfunctions arc examined as part of establishing norms in a society. Then there are belated discoveries—scholars long forgotten who are recalled for a particularly useful discovery (let us say Karl A. Wittfogel on the role of technological domination as the source of Chinese dynasties or others who are remembered for calling into question what may have been unquestioned at the time of creation).

There is a capricious aspect to specific tastes; but there is also a larger rationality that pervades the notion of cultural tradition. There is little purpose to picking sides or choosing among scientific figures of the moment. What remains, the underlying grit of social science as social practice, is the collective response to the struggle between important individuals who have gripped the minds and hearts of people from generation to generation. The transmission of social science is itself a cultural struggle. Most social-science research wanders off in the night; a little of it catches the imagination of more than a single generation; while less stuff still becomes part of the long tradition that links generations to each other.

We should not become involved in false dichotomies and losing battles. By that I mean we must not interpret activities in defense of sociological knowledge as a struggle against modern culture, or varieties of modernity as such. To interpret social science as a struggle between computer literacy and literacies of older sorts of narrative is simply the latest and most dangerous variety of Ludditism. Machine hatred in the form of environmentalism is no less than machine worship in the service of industrialism. Such extremes of theory serve only to weaken the Western tradition to a minority role, a cry in the wilderness of advancing technology. Such a view pits social science against the modern world and, doing such, reduces our disciplines to a fleeting backward glance. And that will not do.

The theorists most admired in sociology, whether Marx, Durkheim, Weber, Simmel, Mannheim, or more recent scholars, were interested in and thus were part of the everyday life of their societies. They studied suicides in France, Protestants in Germany, peasants in Russia, intellectuals in strange lands, strangers in intellectual lands. And this has been even more the case as the social sciences have become linked to a world of military and economic applications. Indeed, what in part we admire most in others is precisely the degree to which the materials of past scholars and scientists were so skillfully manufactured that their words and actions become general principles for our present. In short, theoretical constructs were a by-product of the research process, not a search for grand theory as such.

False antinomies need to be frankly mentioned and just as boldly discarded: the dichotomy between specialization and holistic approaches to the learning experience; the choice between a critical sociology and a scientific sociology; and, for that matter, pure research in any field versus applied-research areas from ceramics to computers, from easel painting to poster design; and, above all, any dichotomy between social studies as conversative and social studies as radical. Uses of knowledge are varied, often uncontrollable. The manufacture of that knowledge cannot, however, be falsified to lean toward one use and away from another. To do so is to falsify science as such and offer scant guarantee of positive social consequence in the bargain.

If such dualisms are allowed to focus our attention and frame our consciousness, then the war is lost, and only skirmishes remain to be fought. A world in which hunger remains, and terror expands, is not likely to want to participate in an exclusively academic dialogue of the deaf—especially when they see it is conducted on high. It is time to substitute a philosophy of "both/and" for one based on "either/or." The premise behind such reasoning is that the risks of eclecticism, while ever present in such a centrist formulation, are far less risky in human terms than demands for theoretical perfections built on dogmatic assumptions.

What we want to carry away from the learning experience is a sense of the ethical or, if you prefer, the moral basis of social science or developmental education. Admittedly, for purposes of this exposition, I treat the two as the same. But left at that, we remain only with an abstraction. For we must also aim to remove a doctrine of intimidation and raw power in favor of verifiable experience and agreed-upon modes of experimentation. We must turn away from the idea that others adjudicate our conditions of life. We must want to adjudicate our own lives. We must spurn the idea that the world is a place of rights and automatic entitlements and accept a world in which there is a balance between rights and obligations, choices and necessities. The task of teaching and learning social science is to facilitate the development and fine tuning of values that help us to adjudicate our lives, to find our own balances. This is a process that enriches our lives by reducing any sense of frustration at not getting everything we want when we want it. It increases joy by showing us a process of careful reflection and self-reflection.

The actual measure of the pervasive decline of social education in the university is the abandonment of ethical discourse to the law.[5] The university has become a battleground of advocates and rights, adjudicated by a battery of lawyers, lawyers manqué, and those who take on the mantle of law in the name of conflict resolution. All of this becomes possible when a concurrence of circumstances takes place—as indeed has happened in higher education today: first, an absolute belief in both rights and obligations, so that the need for judicial process becomes insurmountable; second, a corresponding decline in any sense of obligation, so that few if any are prepared to back away from their demands in the name of a common culture; third, a judicial process capable of becoming intimately involved with university life at every level—from cheating on examination to sexual conduct and misconduct; and fourth, a corresponding reduction in the ability of an administration to conduct campus affairs or, worse, an unwillingness of a faculty to become involved in controversial matters—especially those that might remotely entail litigation.

The worldwide inflation of grades, for instance, is not the cause of a malaise in the learning system but a consequence of the foregoing. The deflation in the value of degrees is, likewise, not a cause of the college malaise but its consequences. The highly legal environment, with its scarcely veiled threat of political mayhem, displaces the ethical environment. As a result, there is a corresponding loss of academic nerve, a decline of serious scholarship into sensitive areas, and, above all, a cynical view toward the process of learning as such. At these levels, the question of social studies is profoundly linked to the requirements of higher education in developmental contexts.

When the dust settles, the contents of the higher learning is not a body of music or a canon of literature inherited from past ages but those disciplines which add nuance and feeling to the hard, flinty tasks of numerical precision or the laws underwriting nature as a whole. Useful learning should provide a realm in which the imagination runs free (but in so doing, still informs our personal lives with meaning and our public life with courage).

The current status of social-scientific education points to powerful gaps, even disparities, in the pattern of evolution. We have experienced in the twentieth century a rapid series of physical and technological changes that are so dazzling as to transform our world. We started the century in horse and buggy, and we end it in supersonic flight. We started it with oil lamps and the bare beginning of electricity, and we end it with superconductivity and chips that store and retrieve information at previously unheard of levels of sophistication.

We still share a moral climate that our parents, and grandparents, would readily recognize. We have continuing animosities of all sorts based on race, religion, and social origin. We have individuals who exhibit greed, avarice, and power unchecked by law. We experience political actions leading to mass slaughter of and mayhem among the innocents. And worse, we try to address these by adjudicating through raw power rather than eradicating through shared moral premises the foundations of such ills. The social sciences are uniquely positioned to reveal the dynamics at work in such polarities.[6]

The developing regions are far from exempt from the problems herein described. Indeed and alas, they epitomize such issues. In the face of such pulls of tradition, we have educational deficiencies that only exacerbate the problem: rote learning, ritualistic ideologies, and presumptions of wisdom based on little else than actualities of classroom recitations. Having served as external examiner for Caribbean universities and having taught in a variety of Third World contexts, I have become convinced that the problem of two cultures, the gap between scientific education and traditional education, is so great that the very selection of the latter over the former invites skepticism of the abilities of the young people involved to manage a modern world. On the other hand, the choice of scientific education is just as frequently perceived as opting out of social and economic problems. In such a context of mistrust generated by the two cultures, the third culture of social-scientific activities, broadly conceived, becomes a source of reconciliation and transformation.

The rise of professionalization and narrow specialization, however lamentable in the abstract, affords a measure of security for those teachers and pupils for whom genuine education and genuine dedication to developmental goals remain alive.[7] The winds of doctrinal change, the revolutionary developments in Russia, China, and, above all, eastern Europe, offer a hope that Third World educational systems will become more sensitive to the need for real learning and less charged with ideological zeal that promises so much and delivers so little. There are such signs: from Mexico to Nigeria to the Philippines is a new wave of concern for giving real meaning to the social in the context of the social scientific—and not confusing scientific studies with ideological posturings. When this wave becomes a flood, then one can address the issue of removing the two cultures, and moving on to a unified educational environment. Until then, however, the duality of cultures offers a measure of support for those who want to assist in projects of national development and, no less, preserve their individual quest for intellectual integrity.

The challenge of the twenty-first century is to narrow the gap between technical progress and moral stagnation. We can best meet this challenge in the social sciences by permitting in the fresh breezes of other times and places, by reading the past's religious literature, its poetry, and its novels and by learning about its heroes, its architecture, sports, foods, occupations, and technology. To fail in this task is to repeat and deepen cycles of dictatorships in fact and double standards in theory. However, in opening wide the gates of learning to others, we must remain true to the basic universal values upon which Western democracy has been built. To transform development into apologetics, into such hoary polarized notions of "Western democracy" versus "Oriental democracy" or "political democracy" versus "economic democracy," is to subvert the foundations of education as such by reducing the learning curve to ideological testing rods. The maintenance of universal standards is a test of supreme importance. Only by so doing can a notion of a common culture endure across boundaries.

This challenge is not one that pertains to advanced countries alone. Developing countries are not exempt from universal claims concerning the moral basis of learning and teaching. In huge portions of the Third World, there is a regression into poverty, mass extermination, fanaticism, despotism, and cultural backwardness. Indeed, for too long a kind of Third World exceptionalism has played havoc with any effort at transforming educational theory into clear developmental goals. Paternalism disguised as concern has only postponed the day of reckoning by confusing the need for exactitude with an inheritance of colonialism. As we enter the final decade of the twentieth century, the worn rhetoric of anticolonialism must finally yield to fundamental concerns that plague the educational environment as a whole. And that means a return to universal concerns, whether they be framed in terms of sovereignty or morality. For the opponents of development, of modernity as a value, also see themselves as operating from universalist premises. So, with S. M. Lipset's end of ideology[8] still only a remote prospect, and Francis Fukuyama's end of history only a remote consequence of Western upheavals,[9] it is best to keep before us the reality of a world in which only a minority accepts the idea of development; and a smaller cluster still, the idea of social science as a corrective to the worst infections of passionate obscurantism.

The social sciences, as unique products of the modern era, are instruments that can help people close the chasm between destructive forms of particularistic credos and ideologies. If a common ground of social science is to prevail, then a shared ethic of discovery and uncovery must be postulated as a mechanism to realize such a goal. The act of creation is what must be preserved in the search for new forms of advancing the higher learning in newly developing nations, no less than in advanced industrial contexts.

In their rich variety, the social sciences offer a common langauge of discourse, logic, and method. They provide a beacon of rational light in a world still largely blanketed by ideological darkness. But the conversion of a specialized series of disciplines into a shared culture, a third culture, is a tough, trying task. To achieve a positive outcome will require a double-edged struggle: against the political barbarians at the gate and against the professional savages who have already gotten inside. The price of success will not be cheap, but the cost of failure—to society at large no less than to the social sciences as such—makes the effort an absolute necessity.

Notes

Introduction

1. Kantrowitz, Barbara, "Sociology's Lonely Crowd," *Newsweek*, vol. 119, no. 5 (Feb. 3, 1992):55–56.

2. Coughlin, Ellen K., "Sociologists Confront Questions About Field's Vitality and Direction," *Chronicle of Higher Education* (Aug. 12, 1992), 5–7 (sect. A).

3. Sowell, Thomas, "Don't Confuse Us with the Numbers," *Forbes*, vol. 149, no. 4 (Feb. 17, 1992):68–69.

4. Marsland, David, *Seeds of Bankruptcy: Sociological Bias Against Business and Freedom* (London: Claridge Press, 1988).

5. Turner, Stephen P., and Dirk Käsler, eds., *Sociology Responds to Fascism* (New York: Routledge, 1992).

6. Hollander, Paul, *Anti-Americanism: Critiques at Home and Abroad, 1965–1990* (New York: Oxford Univ. Press, 1992).

1. The Decomposition of Sociology

1. Collins, Randall, "The Organizational Politics of the ASA," *American Sociologist* 21 (Winter 1990):311.

2. Gollin, Albert E., "Whither the Profession of Sociology?" *American Sociologist* 21 (Winter 1990).

3. Lengyel, Peter, "Elements of Creative Social Science: Towards Greater Authority for the Knowledge Base," *International Social Science Journal* 41 (Nov. 1989).

4. Horowitz, Irving Louis, "Socialist Utopias and Scientific Socialists:

Primary Fanaticisms and Secondary Contradictions," *Sociological Forum* 4 (Mar. 1989).

5. Berger, Brigitte, "The Idea of the University," *Partisan Review* 58 (Spring 1991).

6. Horowitz, Irving Louis, "In Defense of Scientific Autonomy: The Two Cultures Revisited," *Academic Questions* 2 (Winter 1988–89).

7. Cohen, Stanley, *Against Criminology* (New Brunswick: Transaction Publishers, 1988).

8. Hollander, Paul, *Anti-Americanism: Critiques at Home and Abroad, 1965–1990* (New York: Oxford Univ. Press, 1992), esp. 146–213.

9. Oppenheimer, Martin, "Pages from a Journal of the Middle Left," in *Radical Sociologists and the Movement: Experiences, Lessons, and Legacies,* ed. by Martin Oppenheimer, Martin J. Murray, and Rhonda F. Levine (Philadelphia: Temple Univ. Press, 1991), 126–27.

10. Bryden, David P., "It Ain't What They Teach, It's the Way That They Teach It," *Public Interest* 26 (Spring 1991):43.

11. Berger, "The Idea of the University," 321–22.

12. Goheen, Robert F., *Education in United States Schools of International Affairs* (Princeton: Woodrow Wilson School of Public Affairs, 1987), 48.

13. Thompson, Kenneth W., "The Decline of International Studies," *Ethics and International Affairs,* vol. 5 (1991):233–45.

2. Disenthralling Sociology

1. Horowitz, Irving Louis, ed., *The New Sociology: Essays in Social Science and Social Theory* (New York: Oxford Univ. Press, 1964), 3–48.

2. As erstwhile a student of sociological history as Daniel Bell can only come up with the "interpretive turn" in the work of Robert Bellah as indicative of a new, worthwhile vision of the discipline. See the symposium that Bell edited—"New Directions in Modern Thought," *Partisan Review,* vol. 6, no. 2 (1984):215–19.

3. Mommsen, Wolfgang, *Max Weber and German Politics, 1890–1920* (Chicago: Univ. of Chicago Press, 1985).

4. Weber, Max, "Science as a Vocation" (originally delivered as a speech at Munich Univ. in 1918 and first published in 1919), in *From Max Weber: Essays in Sociology,* trans. and ed. by Hans H. Gerth and C. Wright Mills (New York: Oxford Univ. Press, 1946), 129–56.

5. Schuessler, Karl, "Sociology toward the Year 2000," *Society,* vol. 16, no. 5 (1979):31–35.

6. See Wildavsky, Aaron, "No Way without Dictatorship, No Peace without Democracy: Foreign Policy as Domestic Politics," *Social Philosophy & Policy,* vol. 3, no. 1 (1985):176–91.

7. Money, John, and Patricia Tucker, *Sexual Signatures: On Being a Man or a Woman* (Boston: Little, Brown, 1975; also see Money, John,

and Herman Musaph, eds., *Handbook of Sexology* (New York: Elsevier, 1977).

8. Ramey, James W., *Intimate Friendships* (Englewood Cliffs, N.J.: Prentice-Hall, 1976).

9. Levin, Michael, *Feminism and Freedom* (New Brunswick: Transaction Publishers, 1986).

10. Schur, Edwin M., *Crimes without Victims: Deviant Behavior and Public Policy* (Englewood Cliffs, N.J.: Prentice-Hall, 1965). This early pioneering statement was followed by several others in the same vein. *Our Criminal Society: The Social and Legal Sources of Crime in America* (Englewood Cliffs, N.J.: Prentice-Hall, 1969); and *The Politics of Deviance: Stigma, Contexts, and the Uses of Power* (Englewood Cliffs, N.J.: Prentice-Hall, 1979).

11. Taylor, Ian, et al., *The New Criminology* (London: Routledge & Kegan Paul, 1973); and more recently by the same author, *Law and Order: Arguments for Socialism* (Atlantic Highlands, N.J.: Humanities Press, 1981).

12. Black, Donald, "Crime as Social Control," *American Sociological Review*, vol. 48, no. 1 (1983):34–45.

13. Black, Donald, "Common Sense in the Sociology of Law," *American Sociological Review*, vol. 44, no. 1 (1979):18–37.

14. Lears, Jackson, *No Place of Grace: Antimodernism and the Transformation of American Culture, 1880–1920* (New York: Pantheon, 1981), esp. 299–312.

15. Bellah, Robert, et al., *Habits of the Heart: Individualism and Commitment in American Life* (Berkeley: Univ. of California Press, 1983); the same theme is struck, somewhat more cautiously, in his earlier work, *Beyond Belief: Essays on Religion in a Post-Traditional World* (New York: Harper & Row, 1976).

16. Crozier, Michel, *The Trouble with America* (Berkeley: Univ. of California Press, 1984).

17. O'Connor, James, *The Fiscal Crisis of the State* (New York: St. Martin's Press, 1973); and more recently, *Accumulation Crisis* (Oxford: Basil Blackwell, 1984).

18. Roach, Jack L., and Janet K. Roach, "Organizing the Poor: Road to a Dead End," *Social Problems*, vol. 26, no. 2 (1978):160–71.

19. Fox Piven, Frances, and Richard A. Cloward, *Regulating the Poor: The Function of Public Welfare* (New York: Vintage Books, 1971); and by the same authors, "Electoral Instability, Civil Disorder, and Relief Rises," *The American Political Science Review*, vol. 73, no. 4 (1979):1012–19.

20. Domhoff, G. William, *The Bohemian Grove and Other Retreats: A Study in Ruling Class Cohesiveness* (New York: Harper & Row, 1975); and the earlier work by the same author, *The Higher Circles: The Governing Class in America* (New York: Random House, 1970), 251–75.

21. Roy, William G., "The Unfolding of the Interlocking Directorate

Structure of the United States," *American Sociological Review*, vol. 48, no. 2 (1983):248–57.

22. Quadagno, Jill S., "Welfare Capitalism and the Social Security Act of 1935," *American Sociological Review*, vol. 49, no. 5 (1984):632–47.

23. Aminzade, Ronald, "Capitalist Industrialization and Patterns of Industrial Protest," *American Sociological Review*, vol. 49, no. 4 (1984): 437–53.

24. Griffin, Larry J., Michael E. Wallace, and Beth A. Rubin, "Capitalist Resistance to the Organization of Labor Before the New Deal: Why? How? Success?" *American Sociological Review*, vol. 51, no. 2 (1986):164–65.

25. Mushkatel, Alvin, and Khalil Nakhel, "Eminent Domain: Land-Use Planning and the Powerless in the United States and Israel," *Social Problems*, vol. 26, no. 2 (1978):147–59.

26. Mazrui, Ali A., "Zionism and Apartheid: Strange Bedfellows or Natural Allies?" *Alternatives: A Journal of World Policy*, vol. 9, no. 1, (1983):73–97.

27. Mazrui, Ali A., "From the Semites to the Anglo-Saxons: Culture and Civilization in Changing Communications," *Alternatives: A Journal of World Policy*, vol. 11, no. 1 (1986):3–43.

28. Abdel-Malek, Anouar, "Historical Surplus-Value," *Review*, vol. 3, no. 1 (1979):35–44. A more recent statement along the same ideological line is by Seliktar, Ofira, *New Zionism and the Foreign Policy System of Israel* (Carbondale: Southern Illinois Univ. Press, 1986), 268–73.

29. Wallerstein, Immanuel, *The Modern World System: Capitalist Agriculture and the Origins of the European World-Economy in the Sixteenth Century* (New York: Academic Press, 1974); and by the same author, *The Modern World System II: Mercantilism and the Consolidation of the European World Economy, 1600–1750* (New York: Academic Press, 1980).

30. Wallerstein, Immanuel, *Historical Capitalism* (New York: NLB/Schocken, 1983); and *The Politics of the World Economy: The States, the Movements, and the Civilizations* (New York: Cambridge Univ. Press, 1984).

31. Burawoy, Michael, "Between the Labor Process and the State: The Changing Face of Factory Regimes Under Advanced Capitalism," *American Sociological Review*, vol. 48, no. 5 (1983):587–605.

32. See, in particular, Poulantzas, Nicos, *Fascism and Dictatorship* (New York: Schocken Books, 1980); and *State, Power and Socialism* (New York: New Left Books, 1980); and Mandel, Ernest, *From Stalinism to Eurocommunism* (New York: NLB/Schocken, 1978); and *Revolutionary Marxism Today* (New York: NLB/Schocken, 1980).

33. Amid the hardening of the new orthodoxy surrounding "dependency theory," articles are beginning to appear taking the view that one can know the social world through the singular lens of Washington. One recent article notes that contrary to anticapitalist orthodoxy, different positions in the world system are associated with different levels of political democracy,

even after controlling for economic development. Further, authoritarian governments are more typical in the periphery rather than in the semi-periphery. Economic development increases (not decreases) the changes of success for political democracy. See Bollen, Kenneth, "World System Position, Dependency, and Democracy," *American Sociological Review*, vol. 48, no. 4 (1983):468–79.

34. This ideological underpinning has become clear to the brighter critics. Under the rubric of studying social spending in comparative British and U.S. contexts, several creative sociologists have argued unfashionably that welfare is much more than an irreversible by-product of capitalism but no less a process grounded "in the logics of state-building"—in short, in ordinary democratic politics. See Orloff, Ann S., and Theda Skocpol, "Why Not Equal Protection?" *American Sociological Review*, vol. 49, no. 6 (1984):726–50.

35. We must note the extraordinary efforts of a handful of scholars in developing a sociology of Soviet society. In particular, see Hollander, Paul, *Political Pilgrims: Travels of Western Intellectuals to the Soviet Union, China and Cuba* (New York: Oxford Univ. Press, 1981); Simirenko, Alex, *Soviet Sociology: Historical Antecedents and Current Appraisals* (Chicago: Quadrangle Books, 1966); by the same author, *Professionalization of Soviet Society* (New Brunswick: Transaction Publishers, 1982); and Lane, David, *Politics and Society in the USSR* (New York: Random House, 1961).

36. Feuer, Lewis S., "Cultural Détente at a Scientific Congress," *Orbis: A Journal of World Affairs*, vol. 23, no. 1 (1979):115–27.

37. Bartels, Ditta, "It's Good Enough for Science. But Is It Good Enough for Social Action?" *Science, Technology and Human Values*, vol. 10, no. 4 (1985):69–74. For an even blunter attack by an economist, see Yales, Michael D., "South-Africa, Anti-Communism, and Value-Free Science," *The Chronicle of Higher Education* (May 14, 1986), 84.

38. Classical conservatism has too readily identified sociology as a unique enemy of norms of law and with a denuding of moral language. This view has been recently expressed by McClay, Wilfred M., "Liberalism and the Loss of Community," *The Inter-collegiate Review*, vol. 21, no. 3 (1986): 49–51. Just what one is to replace or displace the social sciences with remains as undiscussed as it is undigested. In this, the New Right comes full circle and greets the Old Left, which also saw in sociology a unique enemy of revolutionary norms and laws.

39. Roach, Jack and Janet K. Roach, "Letter on Disenthralling Sociology," *Society*, vol. 24, no. 5 (July–Aug. 1987):3–5.

40. Zaslavskaya, Tatyana, "Perestroika and Sociology," *Social Research*, vol. 55, nos. 1–2 (Fall 1988):267–76; and more recently, "Socialist Voices in the Soviet Union," *Dissent*, vol. 37, no. 2 (Spring 1990): 192–93.

3. Sociology and Subjectivity

1. Longmere, Paul K., *The Invention of George Washington* (Berkeley: Univ. of California Press, 1988), 337.

2. Karnouw, Stanley, *In Our Image: America's Empire in the Philippines* (New York: Random House, 1989), 480.

3. Edelman, Murray J., *Constructing the Political Spectacle* (New York: Random House, 1989), 494.

4. Cohn, Jan, *Creating America: George Horace Lorimer and* The Saturday Evening Post (Pittsburgh: Univ. of Pittsburgh Press, 1989), 326.

5. Greenberg, David F., *The Construction of Homosexuality* (Chicago: Univ. of Chicago Press, 1988), 635.

6. Russett, Cynthia Eagle, *Sexual Science: The Victorian Construction of Womanhood* (Cambridge: Harvard Univ. Press, 1989), 245.

7. Glasser, Ira, "Television and the Construction of Reality," *Etc.,* vol. 31, no. 2. (Spring 1988).

8. Fishman, Mark, *Manufacturing the News* (Austin: Univ. of Texas Press, 1980), 180.

9. Herman, Edward S., and Noam Chomsky, *Manufacturing Consent: The Political Economy and the Mass Media* (New York: Pantheon, 1988), 412.

10. Hill, Michael R., *Mid-American Review of Sociology,* vol. 2, no. 1 (Summer 1984).

11. Pfohl, Stephen, *Death at the Parasite Cafe: Social Science (Fictions) and the Postmodern* (New York: St. Martin's Press, 1992), cap. 59–134

12. Moynihan, Daniel Patrick, "Defining Deviancy Down," *American Scholar,* vol. 62, no. 1 (Winter 1993):17–30.

4. Fascism, Communism, and Social Theory

1. Feuer, Lewis S., "The Social Roots of Einstein's Theory of Relativity," in *Einstein and the Generations of Science* (New Brunswick: Transaction Books 1982), 4–89, which provides an astonishing sense of his versatility in perceiving real disciplines in a fresh way.

2. Feuer, Lewis S., "Cultural Détente at a Scientific Congress: The Secret Report of the Czech Sociologist," ed. and introduced by L. S. Feuer, *Orbis* 23 (Spring 1979):115–27.

3. Feuer, Lewis S., "Introduction," *Marx and Engels: Basic Writings on Politics and Philosophy* (Garden City: Doubleday, 1959), xx–xxi.

4. Feuer, Lewis S., *Ideology and the Ideologists* (New York: Harper & Row, 1975), 120–24.

5. Skidelsky, Ronald, *Oswald Mosley* (New York: Rinehard & Winston, 1975), 465–80.

6. Davis, Kenneth S., *The Hero: Charles A. Lindbergh and the American Dream* (Garden City: Doubleday, 1959), 418–29.

7. Horowitz, Irving Louis, *Winners and Losers: Social and Political Polarities in America* (Durham: Duke Univ. Press, 1984), 179–91.

8. Marcus, Sheldon, *Father Coughlin: The Tumultuous Life of the Priest of the Little Flower* (Boston: Little, Brown, 1973).

9. Brinkley, Alan, *Voices of Protest: Huey Long, Father Coughlin, and the Great Depression* (New York: Alfred A. Knopf, 1982), 281–82.

10. Ribuffo, Leo P., *The Old Christian Right: The Protestant Far Right from the Great Depression to the Cold War* (Philadelphia: Temple Univ. Press, 1973), xiii, 179, 220, *et passim*.

11. Gray, John, "The System of Ruins," *The Times Literary Supplement*, Dec. 30, 1983, pp. 1459–61.

12. Nolte, Ernst, *Marxism, Fascism, Cold War* (Assen, Netherlands: Von Gorcum, 1983), 248.

13. Lenin, Vladimir I., *Left-Wing Communism, An Infantile Disorder: A Popular Essay in Marxian Strategy and Tactics* (New York: International Publishers, 1920), 10–11.

14. Trotsky, Leon, *The Struggle against Fascism in Germany*, with an introduction by Ernest Mandel (New York: Pathfinder Press, 1971), 437–43.

15. Hitchens, Christopher, "Eurofascism: The Wave of the Past," *New Statesman* 231, 18 (1980):567–70.

16. Adorno, Theodore W., "On the Fetish Character in Music and the Regression of Listening" and "Commitment," in *The Essential Frankfurt School Reader*, ed. by Andrew Arato and Eike Gebhartd (New York: Urizen Books, 1978), 270–318.

17. White, Carol, "The Rieman-LaRouche Model: Breakthrough in Thermodynamics," *Fusion* 3, 10 (Aug. 1980):57–66; and Bardwell, Steven, and Uwe Parpart, "Economics Becomes a Science," *Fusion* 2, 9 (July 1970):32–50.

18. Kogan, N., "Fascism as a Political System," in *The Nature of Fascism*, ed. by S. J. Woolf (New York: Random House, 1968), 11–18.

19. Horowitz, Irving Louis, "Preface to the Paperback Edition of Radicalism and the Revolt Against Reason—Then and Now," in *Radicalism and the Revolt Against Reason* (Carbondale: Southern Illinois Univ. Press, 1968).

20. Kahn, H., et al., "NCLS/U.S. Labor Party: Political Chameleon to Right-Wing Spy," *The Public Eye* 1, 1 (Fall 1977):6–22.

21. Abel, Lionel, "Our First Serious Fascist?" *Dissent* 27, 4 (Fall 1980):430–36.

22. Gottlieb, Arthur, "The Dialectics of National Identity: Left-Wing Anti-Semitism and the Arab-Israel Conflict," *Socialist Review* 9, 5 (Sept./Oct. 1979):19–52.

23. Bellent, R. S. Gluss, E. Gordon, H. Kahn, and M. Ryter, *National Caucus of Labor Committees: Brownshirts of the Seventies* (Arlington: Terrorist Information Project, 1977).

24. Beichman, Arnold, "The Myth of American Fascism," in *The Heri-

tage Lectures, no. 7 (Washington, D.C.: Heritage Foundation, 1981):1–22.

25. Arendt, Hannah, *The Origins of Totalitarianism* (New York: Harcourt, Brace, 1966).

26. Halevy, Elie, *The Era of Tyrannies* (New York: New York Univ. Press, 1966).

27. Lichtheim, George, *Collected Essays* (New York: Viking Press, 1973).

28. Reich, Wilhelm, *The Mass Psychology of Fascism* (New York: Farrar, Straus & Giroux, 1979).

29. Talmon, Jacob L., *The Myth of the Nation and the Vision of Revolution* (New Brunswick: Transaction Publishers, 1991).

30. Hollander, Paul, "Sociology and the Collapse of Communism," *Society*, vol. 30, no. 1 (Nov.–Dec. 1992):26–32.

31. Berger, Brigitte, and Peter Berger, "Our Conservatism and Theirs," *Commentary*, vol. 82, no. 4 (Oct. 1986):62–67.

5. From Socialism to Sociology

1. Hinkle, Roscoe, *Founding Theory of American Sociology: 1881–1915* (London: Routledge & Kegan Paul, 1980). See especially the imaginative segments on evolutionism, pp. 183–249.

2. Ross, Dorothy, *The Origins of American Social Science* (New York: Cambridge Univ. Press, 1991).

3. Fleming, Donald, and Bernard Bailyn, *The Intellectual Migration: Europe and America, 1930–1960* (Cambridge: Harvard Univ. Press, 1969).

4. Demerath, N. J. III and Richard A. Peterson, eds., *System, Change, and Conflict* (New York: Free Press, 1967). This compendium contains a range of articles that well reflect this fundamental shift from consensus to conflict, not only within American society but in American sociology.

5. Horowitz, Irving Louis, "Between the Charybdis of Capitalism and the Scylla of Communism: The Emigration of German Social Scientists, 1933–1945," *Social Science History*, vol. 11, no. 2 (Summer 1987):113–38.

6. Lipset, Seymour Martin, "From Socialism to Sociology," in *Sociological Self-Images*, ed. by Irving Louis Horowitz (Beverly Hills: Sage Publications, 1969), 143–76. The British edition is identical in all respects, save for the essay's title change to "Socialism and Sociology" and the publisher and date (Oxford: Pergamon Press, 1970).

7. Sombart, Werner, *The Jews and Modern Capitalism*, with a new introduction by Samuel Z. Klausner (New Brunswick: Transaction Publishers, 1982). All references and citations are to this edition of the Sombart classic.

8. Weber, Max, "The Protestant Sects and the Spirit of Capitalism," in *From Max Weber: Essays in Sociology* (New York: Oxford Univ. Press, 1946), 302–22. Also see Weber, Max, *Ancient Judaism*, ed. by Don Martindale (Glencoe, Ill.: The Free Press, 1967).

9. Wistrich, Robert S., *Antisemitism: The Longest Hatred* (New York: Pantheon, 1992).

10. Mosse, George L., *Masses and Man: Nationalist and Fascist Perceptions of Reality* (New York: Howard Fertig, 1980).

11. Miller, Jack, *Jews in Soviet Culture* (New Brunswick: Transaction Publishers, 1984), esp. 1–22.

12. While it is not uniformly the case that the ethnic factor has been dismissed, with the exception of the remarkable and singular efforts of Alex Inkeles, studies of national character or ethnic structures were and remain at a premium in American sociology. See Inkles, Alex, "National Character Revisited," in *The Tocqueville Review*, ed. by Jesse R. Pitts and Roland Simon, vol. 12 (1990–91):83–117.

13. Arendt, Hannah, *The Origins of Totalitarianism* (New York: World Publishing/Times Mirror, 1958).

14. Johnson, Paul, "Marxism Versus the Jews"; and Wistrich, Robert, "The 'Jewish Question': Left Wing Anti-Zionism in Western Societies," in *Anti-Semitism in the Contemporary World*, ed. by Michael Curtis (Boulder: Westview Press, 1986), 39–60.

15. Weinreich, Max, *Hitler's Professors: The Part of Scholarship in Germany's Crimes Against the Jewish People* (New York: Yiddish Scientific Institute—YIVO, 1946). Despite having been hurriedly completed just after the end of World War II, this work remains the most probing and disturbing account of the acquiescence and participation of the academic social scientists and historians in the Nazi racial doctrines.

16. Hollander, Paul, *Anti-Americanism: Critiques at Home and Abroad, 1965–1990* (New York: Oxford Univ. Press, 1992), 343–55.

17. Reinharz, Shulamit, "Sociologists Working in Israel," *Footnotes* (Apr. 1989), 3–4.

18. Cohen, Stanley, "More on Sociology in Israel," and Elnajjar, Hassan, "A View on Arab Palestine," *Footnotes* (Dec. 1989), 8.

19. Alexander, Edward, "Multiculturalism's Jewish Problem," *Congress Monthly* (Nov.–Dec. 1991), 7–9.

6. Scientific Access and Political Constraints

1. Martino, Joseph P., *Science Funding: Politics and Porkbarrel* (New Brunswick: Transaction Publishers, 1992).

2. Katz, James E., *Congress and National Energy Policy* (New Brunswick: Transaction Publishers, 1984).

3. Larson, Otto N., *Milestones and Millstones: Federal Support for Social Science at the National Science Foundation* (New Brunswick: Transaction Publishers, 1991), 224.

4. Morgan, Harriet P., "Intellectuals as Expert Advisors," *Minerva: A Review of Science, Learning and Policy*, vol. 30, no. 4 (Winter 1992):479–96.

5. Lasswell, Harold D. "The Policy Orientation," in *The Policy Sci-*

ences, ed. by Daniel Lerner and Harold D. Lasswell (Stanford: Stanford Univ. Press, 1951), 3–15.

6. Horowitz, Irving Louis, and James E. Katz, "Brown vs. Board of Education," in *Knowledge for Policy: Improving Education through Research* (New York: Falmer Press and Taylor & Francis, 1991), 237–44.

7. Horowitz, Irving Louis, ed., *The Rise and Fall of Project Camelot: Studies in the Relationship Between Social Science and Practical Politics,* Revised Edition (Cambridge: MIT Press, 1974), 3–47.

7. Public Choice and the Sociological Imagination

1. Coleman, James S., *Foundations of Social Theory* (Cambridge: Harvard Univ. Press, 1990), 1024.

2. Parsons, Talcott, *The Structure of Social Action* (Glencoe, Il.: The Free Press, 1949).

3. Lazarsfeld, Paul F., and Elihu Katz. *The Language of Social Research* (New York: The Free Press, 1965).

4. Mannheim, Karl, *Ideology and Utopia: An Introduction to the Sociology of Knowledge* (New York: Harcourt Brace Jovanovich, 1936).

5. Myrdal, Gunnar, *American Dilemma* (New York: Pantheon, 1975).

6. Schumpeter, Joseph A., *History of Economic Analysis* (New York: Oxford Univ. Press, 1954).

7. Lenski, Gerhard E., *Power and Privilege: A Theory of Social Stratification* (Chapel Hill: Univ. of North Carolina Press, 1984).

8. Nozick, Robert, *Anarchy, State and Utopia* (New York: Basic Books, 1974).

9. Hibbs, Douglas A., Jr., *The American Political Economy: Macroeconomics and Electoral Politics* (Cambridge: Harvard Univ. Press, 1989).

10. Koenig, Rene, *Die Geschichte von Deutscher Soziologie* (Hamburg: Carl Hanser Verlag, 1987), 503.

11. Blau, Peter, *Exchange and Power in Social Life* (New Brunswick: Transaction Publishers, 1986).

12. Parson, Talcott, *Toward a General Theory of Action* (Cambridge: Harvard Univ. Press, 1953).

13. Lasch, Christopher, *The Culture of Narcissism* (New York: Warner Books, 1979).

14. Buchanan, James, *The Political Economy of the Welfare State* (New York: Coronet Books, 1988).

15. Becker, Gary S., *The Economic Approach to Human Behavior* (Chicago: Univ. of Chicago Press, 1978).

16. Burnham, James, *The Managerial Revolution: What Is Happening in the World* (Westport, Conn.: Greenwood, 1972).

17. Simon, Herbert A., and March G. Jones, *Organizations* (New York: John Wiley, 1958).

18. Arrow, Kenneth J., *Applied Research for Social Policy* (Cambridge:

Abt Associates, 1979); and *Individual Choice Under Certainty and Uncertainty* (Cambridge: Harvard Univ. Press, 1984).

19. Herrnstein, Richard J., *Sourcebook in the History of Psychology* (Cambridge: Harvard Univ. Press, 1965).

20. Homans, George C., *Certainties and Doubts* (New Brunswick: Transaction Publishers, 1988); and *Sentiments and Activities* (New Brunswick: Transaction Publishers, 1988).

21. Farmer, Mary K., "On the Need to Make a Better Job of Justifying Rational Choice Thoery," *Rationality and Society*, vol. 4, no. 4 (Oct. 1992).

22. Collins, Randall, "The Rationality of Avoiding Choice," *Rationality and Society*, vol. 5, no. 1 (Jan. 1993).

8. Social Contexts and Cultural Canons

1. Mannheim, Karl, "The Democratization of Culture," in *Essays on the Sociology of Culture* (London: Routledge & Kegan Paul Ltd., 1956), 171–246.

2. Bloom, Allen, *The Closing of the American Mind* (New York: Simon & Schuster, 1987).

3. Schlessinger, Arthur M., Jr., *The Disuniting of America* (Knoxville: White Direct Books, 1991).

4. Anderson, Martin, *Imposters in the Temple: American Intellectuals Are Destroying Our Universities and Cheating Our Students of Their Future* (New York: Simon & Schuster, 1992).

5. Sykes, Charles J., *ProfScam: Professors and the Demise of Higher Education* (Chicago: Regency Gateway, 1988).

6. de Beauvoir, Simon, *The Second Sex* (New York: Alfred A. Knopf, 1953).

7. Franklin, John Hope, and August Meier, *Black Leaders of the Twentieth Century* (Champaign/Urbana: Univ. of Illinois Press, 1982), Introduction.

9. Reconstructing the Social Sciences

1. Data on membership of the APSA is regularly supplied by *PS: Political Science and Politics*, and on the ASA by *Footnotes*, its official newsletter. Variations do occur over time—but the long run tends to indicate little variation from that last cited in the text.

2. Horowitz, Irving Louis, "Is Social Science a God that Failed?" *Public Opinion*, vol. 4, no. 5 (Oct.–Nov. 1981):11–12.

3. Titmuss, Richard M., *Problems of Social Policy* (New York: Kraus Publishers International, 1976).

4. Mishan, Ezra, *What Political Economy Is All About: An Exposition and Critique* (Cambridge: Cambridge Univ. Press, 1982), 256; and *Economic*

Myths and the Mythology of Economics (Atlantic Highlands, N.J.: Humanities Publishers, 1986), 268.

5. Weber, Max, "Science as a Vocation" and "Politics as a Vocation," in *From Max Weber: Essays in Sociology*, ed. by H. H. Gerth and C. W. Mills (New York: Oxford Univ. Press, 1946).

6. Johnson, Guy B., "Southern Offensive" and "Memoir" in *The War Within: From Victorian to Modernist Thought in the South, 1919–1945* ed. by Daniel Joseph Singal (Chapel Hill: Univ. of North Carolina Press, 1982), 326.

7. Greenberg, John L., and Judith R. Goodstein, "Theodore von Karman and Applied Mathematics in America," *Science*, vol. 222, no. 4630 (Dec. 23, 1983):1300–1304.

8. Nisbet, Robert A., *The Sociological Tradition* (New York: Basic Books, 1966), 318.

10. Human Life, Political Domination, and Social Science

1. Vinocur, John, "In West Berlin: A New Curtain Rises on Auschwitz," *New York Times*, Apr. 7, 1980, p. 2.

2. Bauer, Yehuda, *The Holocaust in Historical Perspective* (Seattle: Univ. of Washington Press, 1978), 39–49.

3. Wiesel, Elie, *Legends of Our Time* (New York, 1968), 6. For an analysis of this vision, see Des Pres, Terrence, "The Authority of Silence in Elie Wiesel's Art," in *Confronting the Holocaust: The Impact of Elie Wiesel*, ed. by Alvin H. Rosenfield and Irving Greenberg (Bloomington: Univ. of Indiana Press, 1978), 49–57.

4. Donaldson, Peter J., "In Cambodia, A Holocaust," *New York Times*, Apr. 22, 1980, p. 17.

5. Fackenheim, Emil L., "What the Holocaust Was Not," *Face to Face* (an interreligious bulletin issued by the Anti-Defamation League of B'nai B'rith) 7 (Winter 1980): 8–9. This set of propositions is derived from Fackenheim's Foreword to Bauer, Yehuda, *The Jewish Emergence from Powerlessness* (Toronto: Univ. of Toronto Press, 1979).

6. Kuper, Leo, *Genocide: Its Political Use in the Twentieth Century* (New Haven: Yale Univ. Press, 1982).

7. For a fuller version of what has become the dominant and most widely respected Jewish viewpoint on the Holocaust, see Fackenheim, Emil L., *God's Presence in History* (New York: Macmillan, 1972), 70–73.

8. Lifton, Robert J., "The Concept of the Survivor," in *Survivors, Victims, and Perpetrators: Essays on the Nazi Holocaust*, ed. by Joel E. Dimsdale (Washington, D.C.: Univ. Press of America, 1980), 113–26.

9. Literature on the Armenian subjugation is uneven, and only now are scholars facing up to the Herculean research tasks involved. An excellent compendium of available materials for 1915–23 is contained in Hovan-

nisian, Richard G., *The Armenian Holocaust* (Cambridge, Mass.: Harvard Univ. Press, 1980).

10. Pawelczynska, Anna, *Values and Violence in Auschwitz* (Berkeley: Univ. of Calif. Press, 1979), 101–5.

11. Hilberg, Raul, "German Railroads/Jewish Souls," *Transaction/ SOCIETY* 14 (Nov.–Dec. 1976):60–74. For a general introduction to this subject see *Captured Germans and Related Records: A National Archives Conference*, ed. by Robert Wolfe (Athens, Ohio: Ohio Univ. Press, 1974).

12. Mosse, George L., *Toward the Final Solution: A History of European Racism* (New York: Howard Fertig Publishers, 1978), 226–27.

13. Hausner, Gideon, "Six Million Accusers," in *The Jew in the Modern World: A Documentary History*, ed. by Paul R. Mendes-Flohr and Jehuda Reinharz (New York: Oxford Univ. Press, 1980), 521–23.

14. Horowitz, Irving Louis, *Foundations in Political Sociology* (New York: Harper & Row, 1972), 245–46. See in this connection Niewyk, D. L., *Socialist Anti-Semite, and Jew: German Social Democracy Confronts the Problems of Anti-Semitism* (Baton Rouge: Univ. of Louisiana Press, 1971).

15. For a full discussion of the orthodox (minority) viewpoint on the Holocaust in the context of Yeshiva life, see Helmreich, William, "Understanding the Holocaust: The Yeshiva View," *Transaction/SOCIETY* 18 (1980–81).

16. Butz, Arthur B., *The Hoax of the Twentieth Century* (Torrance, Calif.: The Institute for Historical Review, 1976).

17. See App, Austin J., "The 'Holocaust' Put in Perspective," *Journal of Historical Review* 1 (Spring 1980):43–58.

18. Wever, Charles E., "German History from a New Perspective," *Journal of Historical Review* 1 (Spring 1980):81–82.

19. The most authoritative estimate of the number of Jews killed by the Nazis—5,978,000 out of a prewar population of 8,301,000, or 72 percent—is contained in Poliakov, Leon, and Josef Wulf, eds., *Das Dritte Reich und die Juden: Dokumente und Aufsaetze* (Berlin: Arani Verlag, 1955), 229.

20. For an articulate statement of legal and social issues at an individual level which have direct relevance to our discussion at the collective level, see Bereday, George Z. F., "The Right to Live and the Right to Die: Some Considerations of Law and Society in America," *Man and Medicine* 4, no. 4 (1979):233–56.

21. Melson, Robert F., *Revolution and Genocide: On the Origins of the Armenian Genocide and the Holocaust* (Chicago: Univ. of Chicago Press, 1992), 363.

11. Policy Research in a Post-Sociological Environment

1. Gowa, Joanne, "Bipolarity, Multipolarity, and Free Trade," *American Political Science Review*, vol. 83, no. 4 (1989):1245–56.

2. Hansen, Wendy L., "The International Trade Commission and the Politics of Protectionism," *American Political Science Review*, vol. 84, no. 1 (1990):21–46.

3. Sorensen, Theodore C., "Rethinking National Security," *Foreign Affairs*, vol. 69, no. 3 (1990):1–18.

4. Podhoretz, Norman, "Enter the Peace Party," *Commentary*, vol. 91, no. 1 (1991):17–22; and Maynes, Charles William, "A Necessary War?" *Foreign Policy*, 82 (Spring 1991):159–77.

5. Bowsher, Charles A., *Comptroller General's 1990 Annual Report* (Washington, D.C.: United States General Accounting Office, 1991).

6. Wooten, F. Thomas, *Research Triangle Institute Report* (Research Triangle Park: RTI, 1991).

7. Holsti, Ole R., and James N. Rosenau, "The Emerging U.S. Consensus on Foreign Policy," *Orbis: A Journal of World Affairs*, vol. 344, no. 4 (1990):579–94.

8. Clark, John G., *The Political Economy of World Energy: A Twentieth-Century Perspective* (Chapel Hill: Univ. of North Carolina Press, 1990), 376.

9. Hyland, William G., "Downgrade Foreign Policy," *New York Times*, May 20, 1991, p. A15. While the Hyland statement puts a different spin on the subject of foreign and domestic priorities, his discomfort with the present positivist mood is noteworthy.

10. Smith, William C., "Democracy, Distributional Conflicts and Macroeconomic Policymaking in Argentina, 1983–89," *Journal of Interamerican Studies and World Affairs*, vol. 32, no. 2 (1990):1–42.

11. Teichman, Judith A., *Policymaking in Mexico: From Boom to Crisis* (Boston: Allen & Unwin, 1989).

12. Kant, Immanuel, *Critique of Pure Reason* (1781), trans. by Norman Kemp Smith (London: Macmillan Publishers, 1929), 56–57. For a critical, but useful analysis of the Kantian and Hegelian visions, see Kaufman, Walter, *Goethe, Kant and Hegel: Discovering the Mind* (New Brunswick: Transaction Publishers, 1991), 82–156.

12. Prediction and Paradox in Society

1. See, for example, Mirowski, Philip, *More Heat Than Light: Economics as Social Physics, Physics as Nature's Economics* (New York: Cambridge Univ. Press, 1992).

2. Park, Robert E., *Human Communities: The City and Human Ecology* (New York: Free Press, 1952).

3. Lynd, Robert S., and Helen Merrell Lynd, *Middletown: A Study in Contemporary American Culture* (New York: Harcourt, 1930); and *Middletown in Transition: A Study in Cultural Conflicts* (New York: Harcourt, 1937).

4. Powell, Elwin H., *The Design of Discord: Studies of Anomie* (New York: Oxford Univ. Press, 1970).

5. "Woman in Ruling on Mentally Ill Drowns," *New York Times,* Dec. 1, 1990.

6. Faison, Seith, Jr., "Teen-Age Girl Found Chained in Bronx Residence," *New York Times,* Sept. 15, 1991.

7. Bishop, Katherine, "Photos of Nude Children Spark Obscenity Debate," *New York Times,* July 23, 1990.

8. Sack, Kevin, "School Mural: Sociology or Religion?" *New York Times,* May 1, 1990.

9. Greenhouse, Linda, "Symbols Clashing in Debate Over Flag," *New York Times,* May 21, 1991.

10. Treaster, Joseph B., "Is the Fight on Drugs Eroding Civil Rights?" *New York Times,* July 9, 1990.

11. Kolbert, Elizabeth, "New York's Top Court Allows Random Drug Tests of Guards," *New York Times,* May 9, 1990.

12. Lewin, Tamar, "Children as Neighbors? Elderly Are Worried," *New York Times,* Nov. 28, 1989.

13. Chafets, Ze'ev, *Devil's Night: And Other True Tales of Detroit* (New York: Random House, 1990). Extracted in *The New York Times Magazine* (July 29, 1990), 32–51.

14. Mydans, Seth, "University's Choice: Stars or Squirrels," *New York Times,* May 15, 1990.

15. Wald, Matthew L., "Gasahol May Cut Monoxide but Raise Smog," *New York Times,* May 12, 1990.

16. Verhovek, Sam Howe, "Whose Law Applies When Lawlessness Rules on Indian Land?" *New York Times,* April 29, 1990; and Lorch, Donatella, "Behind Violence, Tensions Roil Mohawks," *New York Times,* Apr. 30, 1990.

17. Fein, Esther B., "With Easing of Soviet Police State, Crime Against Foreigners Is Rising," *New York Times,* May 6, 1990.

18. Rothenberg, Randall, "Noriega Tape Case Reviving Clash Over First Amendment," *New York Times,* Nov. 14, 1990.

19. Weber, Max, "Politics as a Vocation," in *From Max Weber: Essays in Sociology,* ed. by Hans H. Berth and C. Wright Mills (New York: Oxford Univ. Press, 1946), 117.

13. Freedom, Planning, and the Moral Order

1. Mannheim, Karl, *Freedom, Power, and Democratic Planning* (New York: Oxford Univ. Press, 1950).

2. Horowitz, Irving Louis, "Social Planning and Social Science: Historical Continuities and Comparative Discontinuities," in *Planning Theory in the 1980s,* ed. by Robert W. Burchell and George Sternlieb (New Brunswick: Center for Urban Policy Research, 1979), 41–68.

3. The literature on models of capitalist development is immense, but quite different perspectives are outlined in Novak, Michael, *The Spirit of Democratic Capitalism* (New York: Basic Books, 1979); Galbraith, John Kenneth, *The New Industrial State* (Boston: Houghton Mifflin, 1967); and Macpherson, C. B., *Democratic Theory: Essays in Retrieval* (New York: Oxford Univ. Press, 1973).

4. Keynes, John Maynard, *The General Theory of Employment, Interest and Money* (London: Macmillan, 1936). The best critique of Keynes's classic work is still Hicks, John, *The Crisis of Keynesian Economics* (Oxford: Basil Blackwell Ltd., 1974).

5. Mannheim, Karl, *Man and Society in an Age of Reconstruction* (London: Routledge & Kegan Paul, 1940).

6. Laski, Harold, *The State in Theory and Practice* (New York: Viking Press, 1935); and by the same author, *The Rise of European Liberalism: An Essay in Interpretation* (London: Allen & Unwin, 1936).

7. Keynes, John Maynard, "The End of Laissez-Faire," in *The Collected Writings of John Maynard Keynes* (New York: St. Martin's Press, 1972).

8. Wooton, Barbara A., *Freedom Under Planning* (Chapel Hill: Univ. of North Carolina Press, 1945); and Beveridge, William Henry, *Full Employment in a Free Society* (New York: W.W. Norton, 1945).

9. Myrdal, Gunnar, *Against the Stream: Critical Essays on Economics* (New York: Pantheon, 1972); and by the same author, *The Political Element in the Development of Economic Theory* (New Brunswick: Transaction Publishers, 1990).

10. Mallock, William Hurrell, *A Critical Examination of Socialism* (New Brunswick: Transaction Publishers, 1989).

11. Lippmann, Walter, *The Good Society* (Boston: Little, Brown, 1937).

12. Friedman, Milton, *Capitalism and Freedom* (Chicago: Univ. of Chicago Press, 1962); and with Rose Friedman, *The Freedom to Choose* (Harmondsworth, Middlesex: Penguin/Viking, 1980).

13. Hayek, A. Friedrich, *Capitalism and the Historians* (London: Routledge & Kegan Paul, 1962); and *The Constitution of Liberty* (London: Routledge & Kegan Paul, 1960).

14. Douglas, Jack D., *The Myth of the Welfare State* (New Brunswick: Transaction Publishers, 1989).

15. Jansson, Bruce S., *The Reluctant Welfare State: A History of American Social Welfare Policies* (Belmont: Wadsworth Publishing Company, 1988); and an earlier review of this same subject by Wilensky, Harold, and Charles Lebeaux, *Industrial Society and Social Welfare* (New York: Free Press, 1965).

16. Rawls, John, *A Theory of Justice* (Oxford: Oxford Univ. Press, 1972). See also, Kukathas, Chandran, and Philip Petit, *Rawls: A Theory of Justice and its Critics* (Stanford: Stanford Univ. Press, 1990).

17. Nozick, Robert, *Anarchy, State and Utopia* (Oxford: Basil Black-

well, 1974); and his more recent effort, *The Examined Life* (New York: Simon & Schuster, 1989). See also, Wolff, Jonathan, *Robert Nozick: Property, Justice and the Minimal State* (Stanford: Stanford Univ. Press, 1991).

18. Buchanan, James M., *The Power to Tax: Analytical Foundations of a Fiscal Constitution* (Cambridge: Cambridge Univ. Press, 1980); for two surveys of public choice doctrines, see Buchanan, James M., *Theory of Public Choice: Political Applications of Economics* (Ann Arbor: Univ. of Michigan Press, 1972); and Gwartney, James D., and Richard E. Wagner, eds., *Public Choice and Constitutional Economics* (Greenwich: JAI Press, 1988).

14. Social Disputations and Moral Implications

1. Horowitz, Irving Louis, "Moral Development, Authoritarian Distemper, and the Democratic Persuasion," in *Moral Development and Politics*, ed. by Richard W. Wilson and Gordon J. Schochet (New York: Praeger, 1980), 5–21.

2. Wilson, John, *Discipline and Moral Education: A Survey of Public Opinion and Understanding* (Windsor: Nelson Publishing Co., 1981). Far more revealing still is Wilson's reply to D. W. Livingstone's review of this work, "Reply to Livingstone's 'On Moral Philsophy and Educational Practice,'" *Curriculum Inquiry*, vol. 13, no. 1 (Spring 1983):95–97.

3. Talmon, Jacob L., *The Myth of the Nation and the Vision of Revolution* (Berkeley: Univ. of Calif. Press, 1981).

4. Arendt, Hannah, *The Origins of Totalitarianism* (New York: Harcourt, Brace, 1966).

5. Halevy, Elie, *The Era of Tyrannies* (New York: New York Univ. Press, 1966).

6. Lichtheim, George, *Collected Essays* (New York: Viking Press, 1973).

7. Lynd, Robert S., and Helen M. Lynd, *Middletown: A Study in Contemporary American Culture* (New York: Harcourt, 1929); and *Middletown in Transition: A Study in Cultural Conflicts* (New York: Harcourt, 1937). My discussion of the Lynds' work is contained in the *International Encyclopedia of the Social Sciences*, ed. by David L. Sills, vol. 18 (New York: Free Press, 1971):471–77.

8. See, for example, Davis, Kingsley, "A Conceptual Analysis of Stratification," *American Sociological Review*, vol. 7 (1941):309–21; Davis Kingsley, and Wilbert E. Moore, "Some Principles of Stratification," *American Sociological Review*, vol. 10 (1945):242–49; and for a rebuttal, see Timin, Melvin, "Some Principles of Stratification: A Critical Review," *American Sociological Review*, vol. 18 (1953):387–94. For a fuller discussion of these issues in "classic" form, see Bendix Reinhard, and Seymour Martin Lipset, *Class, Status, and Power: Social Stratification in Comparative Perspective*, Second Edition (New York: Free Press, 1966).

9. Schattschneider, E. E., *The Semisovereign People: A Realist's View*

of Democracy in America (New York: Holt, Reinhart & Winston, 1960), 141.

10. Truman, David B., *The Governmental Process: Political Interests and Public Opinion* (New York: Alfred Knopf, 1951).

11. Black, Donald, "Crime as Social Control," *American Sociological Review*, vol. 48 (Feb. 1983):34–454.

12. Benedict, Ruth, *Patterns of Culture* (Boston: Houghton Mifflin, 1934); and Kroeber, A. L., ed., *Anthropology Today: An Encyclopedic Inventory* (Chicago: Univ. of Chicago Press, 1953).

13. Malinowski, Bronisław, "The Primitive Economics of the Trobriand Islanders," *Economic Journal*, vol. 31 (1921):1–16.

14. Tax, Sol, ed., *The People vs. the System: A Dialogue in Urban Conflict* (Chicago: Univ. of Chicago Press, 1968).

15. Von Hayek, Friedrich A., *The Constitution of Liberty* (London: Routledge & Kegan Paul, 1960); and Friedman, Milton, *Capitalism and Freedom* (Chicago: Univ. of Chicago Press, 1962).

16. See, for example, Keynes, John Maynard, *General Theory of Employment, Interest and Money* (London: Macmillan, 1936); and Lange, Oskar, "On the Economic Theory of Socialism" (in two parts), *Review of Economic Studies*, vol. 4 (1936):53–71; vol. 5 (1937):123–42.

17. Robinson, Joan, *The Accumulation of Capital* (London: Macmillan, 1956); and *Economic Heresies: Some Old-Fashioned Questions in Economic Theory* (London: Macmillan, 1971).

18. Lerner, Abba P., *The Economics of Employment* (New York: McGraw-Hill, 1951); and *Everybody's Business* (East Lansing: Michigan State Univ. Press, 1961).

19. Fromm, Erich, *The Art of Loving* (New York: Harper & Row, 1956); and *The Sane Society* (New York: Holt, Reinhart & Winston, 1955).

20. Rogers, Carl B., *On Becoming a Person: A Therapist's View of Psychotherapy* (Boston: Houghton Mifflin, 1961); and *Becoming Partners: Marriage and Its Alternatives* (New York: Delacorte, 1972).

21. Walzer, Michael, *Obligations: Essays on Disobedience, War and Citizenship* (Cambridge: Harvard Univ. Press, 1982).

22. Goertzel, Ted George, and Albert J. Szymanski, *Sociology: Class, Consciousness, and Contradictions* (New York: Van Nostrand, 1979); and Goertzel, Ted George, *Political Sociology* (Chicago: Rand McNally, 1976).

23. Inkeles, Alex, *Social Change in Soviet Russia* (Cambridge: Harvard Univ. Press, 1968); and *The Soviet Citizen: Daily Life in a Totalitarian Society*, with Raymond Bauer (Cambridge: Harvard Univ. Press, 1959).

24. Lipset, Seymour Martin, "Economic Equality and Social Class," in *Equity, Income, and Policy*, ed. by Irving Louis Horowitz (New York: Praeger, 1977), 278–86.

25. Hollander, Paul, *Soviet and American Society: A Comparison* (New York: Oxford Univ. Press, 1973); and *Political Pilgrims* (New York: Oxford Univ. Press, 1981).

26. Dahl, Robert A., *After the Revolution? Authority in a Good Society* (New Haven: Yale Univ. Press, 1970); and Lane, David, *Politics and Society in the USSR* (New York: Random House, 1971).

27. Carlson, Allan C., "Sex According to Social Science," *Policy Review*, no. 20 (Spring 1982): 115–39; and Goldberg, Steven, "Is Homosexuality Normal?" *Policy Review*, no. 2 (Summer 1982):119–38.

28. Carrington, Frank G., *The Victims* (New Rochelle: Arlington House, 1975); Viano, Emilio C., "Victimology and its Pioneers," *Victimology: An International Journal*, vol. 1, no. 2 (1976):189–92; Drapkin, Israel and Emilio C. Viano, *Victimology: A New Focus* (Boston: D.C. Heath, 1974–75); and Viano, Emilio, *Victims and Society* (1976).

29. Bell, Alan P., and Martin S. Weinberg, *Homosexualities* (New York: Basic Books, 1980); and Weinberg, Martin S., and Colin J. Williams, *Male Homosexuals* (New York: Oxford Univ. Press, 1974).

30. Humphreys, Laud, *Out of the Closets: The Sociology of Homosexual Liberation* (Englewood Cliffs, N.J.: Prentice Hall, 1972).

31. Goldberg, Steven, *The Inevitability of Patriarchy* (London: M. T. Smith, Ltd., 1977); and by the same author, "Is Homosexuality Normal?" *Policy Review*, no. 21 (Summer 1982):119–38.

32. Lorenz, Konrad, *On Aggression* (New York: Harcourt, Brace, 1974); and *Behind the Mirror* (New York: Harcourt, Brace, 1977).

33. Berman, Marshal, *The Politics of Authenticity: Radical Individualism and the Emergence of Modern Society* (New York: Atheneum, 1970).

34. Flatham, Richard, *Political Obligation* (New York: Atheneum, 1972).

35. Pocock, John G. A., *Politics, Language and Time: Essays on Political Thought and History* (New York: Atheneum, 1971).

36. Jaffa, Harry V., *Equality and Liberty: Theory and Practice in American Politics* (New York: Oxford Univ. Press, 1965), 209–29.

37. Page, Benjamin, *Who Gets What from Government* (Berkeley: Univ. of Calif. Press, 1983).

38. Polsby, Nelson W., *The Consequences of Party Reform* (New York: Oxford Univ. Press, 1983).

39. Mead, Margaret, *Coming of Age in Samoa: A Psychological Study of Primitive Youth for Western Civilization* (New York: William Morrow, 1928).

40. Freeman, Derek, *Margaret Mead and Samoa: The Making and Unmaking of an Anthropological Myth* (Cambridge: Harvard Univ. Press, 1983), 281–302.

41. Mayer, Ernest, *The Growth of Biological Thought: Diversity, Evolution and Inheritance* (Cambridge: Harvard Univ. Press, 1982).

42. Tingergen, Nickolaas, *The Animal in Its World: Explorations of an Ethologist, 1932–1972* (London: Allen & Unwin, 1972–73).

43. Mishan, E. J., *Introduction to Normative Economics* (New York:

Oxford Univ. Press, 1981); and *Cost-Benefit Analysis*, Expanded Edition (New York: Praeger, 1976).

44. Liberman, Evsei Grigor'evich, *Economic Methods on the Effectiveness of Production* (White Plains, N.Y.: International Arts and Science Press, 1972); see also Gardner, Kent, *The Formulation of Economic Policy in the Soviet Union: The Political Evolution of Libermanism* (Chapel Hill: Univ. of North Carolina Press, 1972).

45. Mises, Ludwig von, *Theory and History: An Interpretation of Social and Economic Evolution* (New Haven: Yale Univ. Press, 1957); and *Socialism: An Economic and Sociological Analysis* (New Haven: Yale Univ. Press, 1951).

46. Stigler, George J., *The Citizen and the State: Essays on Regulation* (Chicago: Univ. of Chicago Press, 1975).

47. Myrdal, Gunnar, *The Challenge of World Poverty: A World Anti-Poverty Program in Outline* (New York: Pantheon, 1970); and *Against the Stream: Critical Essays on Economics* (New York: Pantheon, 1973).

48. Featherman, David L., "Social Mobility: Opportunities Are Expanding"; and Willhelm, Sidney M., "Social Mobility: Opportunities are Diminishing," *Transaction/SOCIETY*, vol. 16, no. 3 (Mar. 1979):4–17.

49. Edel, Abraham, *Relating Humanities and Social Thought: Science, Ideology and Value*, Vol. Four (New Brunswick: Transaction Publishers, 1990); and also his seminal work, *Method in Ethical Theory* (Indianapolis: Bobbs Merrill, 1963), 273–79.

15. Social Science and the Great Tradition

1. Hackenberg, Michael, ed., *Getting the Books Out: Papers of the Chicago Conference on the Book in 19th-Century America* (Washington, D.C.: The Center for the Book, Library of Congress, 1987), 197.

2. Morton, Herbert C., et al., *Writings on Scholarly Communication* (Washington, D.C.: Office of Scholarly Communication and Technology, American Council of Learned Societies, 1988), 152.

3. Tebbel, John, *Between Covers: The Rise and Transformation of Book Publishing in America* (New York: Oxford Univ. Press, 1986), 512.

4. Joyce, Donald Franklin, "Reflections on the Changing Publishing Objectives of Secular Black Book Publishers, 1900–1986," in *Reading in America: Literature and Social History*, ed. by Cathy N. Davidson (Baltimore: Johns Hopkins Univ. Press, 1989), 226–39.

5. West, James L. W., III, *American Authors and the Literary Marketplace since 1900* (Philadelphia: Univ. of Pennsylvania Press, 1988), 172.

16. Social Science as the Third Culture

1. Horowitz, Irving Louis, "In Defense of Scientific Autonomy: The Two Cultures Revisited," *Academic Questions*, vol. 2, no. 1 (Winter 1988–89):22–27.

2. Machlup, Fritz, "Are The Social Sciences Inferior?" *Society*, vol. 25, no. 4 (May/June 1988):57–66.

3. Hook, Sidney, "The Politics of Curriculum Building," *Measure: University Centers for Rational Alternatives*, no. 81 (Jan. 1989):1–12.

4. Rummel, R. J., *Lethal Politics: Soviet Genocide and Mass Murder Since 1917* (New Brunswick: Transaction Publishers, 1990).

5. Glazer, Nathan, "Regulating Business and the Universities," *The Public Interest*, no. 56 (Summer 1979):43–62.

6. Shils, Eward A., "Science in the Public Arena," *The American Scholar*, vol. 56, no. 2 (Spring 1987):185–204.

7. MacRae, Duncan, Jr., "Professional Knowledge for Policy Discourse," *Knowledge in Society*, vol. 1, no. 3 (Fall 1988):6–24.

8. Lipset, Seymour Martin, "Neoconservatism: Myth and Reality," *Society* vol. 25, no. 5 (July–Aug. 1988):29–37.

9. Fukuyama, Francis, "The End of History?" *The National Interest*, (Summer 1989):3–18.

Name Index

Subject Index